T0291706

Low Power to the People

Inside Technology Series

edited by Wiebe E. Bijker, W. Bernard Carlson, and Trevor Pinch

Christina Dunbar-Hester, *Low Power to the People: Pirates, Protest, and Politics in FM Radio Activism*

Eden Medina, Ivan da Costa Marques, and Christina Holmes, editors, *Beyond Imported Magic: Essays on Science, Technology, and Society in Latin America*

Anique Hommels, Jessica Mesman, and Wiebe E. Bijker, editors, *Vulnerability in Technological Cultures: New Directions in Research and Governance*

Amit Prasad, *Imperial Technoscience: Transnational Histories of MRI in the United States, Britain, and India*

Charis Thompson, *Good Science: The Ethical Choreography of Stem Cell Research*

Tarleton Gillespie, Pablo J. Boczkowski, and Kirsten A. Foot, editors, *Media Technologies: Essays on Communication, Materiality, and Society*

Catelijne Coopmans, Janet Vertesi, Michael Lynch, and Steve Woolgar, editors, *Representation in Scientific Practice Revisited*

Rebecca Slayton, *Arguments that Count: Physics, Computing, and Missile Defense, 1949–2012*

Stathis Arapostathis and Graeme Gooday, *Patently Contestable: Electrical Technologies and Inventor Identities on Trial in Britain*

Jens Lachmund, *Greening Berlin: The Co- Production of Science, Politics, and Urban Nature*

Chikako Takeshita, *The Global Biopolitics of the IUD: How Science Constructs Contraceptive Users and Women's Bodies*

Cyrus C. M. Mody, *Instrumental Community: Probe Microscopy and the Path to Nanotechnology*

Morana Alač, *Handling Digital Brains: A Laboratory Study of Multimodal Semiotic Interaction in the Age of Computers*

Gabrielle Hecht, editor, *Entangled Geographies: Empire and Technopolitics in the Global Cold War*

Michael E. Gorman, editor, *Trading Zones and Interactional Expertise: Creating New Kinds of Collaboration*

Matthias Gross, *Ignorance and Surprise: Science, Society, and Ecological Design*

Andrew Feenberg, *Between Reason and Experience: Essays in Technology and Modernity*

Wiebe E. Bijker, Roland Bal, and Ruud Hendricks, *The Paradox of Scientific Authority: The Role of Scientific Advice in Democracies*

Park Doing, *Velvet Revolution at the Synchrotron: Biology, Physics, and Change in Science*

Gabrielle Hecht, *The Radiance of France: Nuclear Power and National Identity after World War II*

Richard Rottenburg, *Far-Fetched Facts: A Parable of Development Aid*

Michel Callon, Pierre Lascoumes, and Yannick Barthe, *Acting in an Uncertain World: An Essay on Technical Democracy*

Ruth Oldenziel and Karin Zachmann, editors, *Cold War Kitchen: Americanization, Technology, and European Users*

Deborah G. Johnson and Jameson W. Wetmore, editors, *Technology and Society: Building Our Sociotechnical Future*

Trevor Pinch and Richard Swedberg, editors, *Living in a Material World: Economic Sociology Meets Science and Technology Studies*

Christopher R. Henke, *Cultivating Science, Harvesting Power: Science and Industrial Agriculture in California*

Helga Nowotny, *Insatiable Curiosity: Innovation in a Fragile Future*

Karin Bijsterveld, *Mechanical Sound: Technology, Culture, and Public Problems of Noise in the Twentieth Century*

Peter D. Norton, *Fighting Traffic: The Dawn of the Motor Age in the American City*

Joshua M. Greenberg, *From Betamax to Blockbuster: Video Stores tand the Invention of Movies on Video*

Mikael Hård and Thomas J. Misa, editors, *Urban Machinery: Inside Modern European Cities*

Christine Hine, *Systematics as Cyberscience: Computers, Change, and Continuity in Science*

Wesley Shrum, Joel Genuth, and Ivan Chompalov, *Structures of Scientific Collaboration*

Shobita Parthasarathy, *Building Genetic Medicine: Breast Cancer, Technology, and the Comparative Politics of Health Care*

Kristen Haring, *Ham Radio's Technical Culture*

Atsushi Akera, *Calculating a Natural World: Scientists, Engineers and Computers during the Rise of U.S. Cold War Research*

Donald MacKenzie, *An Engine, Not a Camera: How Financial Models Shape Markets*

Geoffrey C. Bowker, *Memory Practices in the Sciences*

Christophe Lécuyer, *Making Silicon Valley: Innovation and the Growth of High Tech, 1930–1970*

http://mitpress.mit.edu/books/series/inside-technology

Low Power to the People

Pirates, Protest, and Politics in FM Radio Activism

Christina Dunbar-Hester

The MIT Press
Cambridge, Massachusetts
London, England

This book was set in Stone Sans and Stone Serif by the MIT Press.

Library of Congress Cataloging-in-Publication Data

Dunbar-Hester, Christina, 1976-
Low power to the people : pirates, protest, and politics in FM radio activism / Christina Dunbar-Hester.
pages cm. — (Inside technology)
Includes bibliographical references and index.
ISBN 978-0-262-02812-7 (hardcover : alk. paper) — 978-0-262-53476-5 (paperback)
1. Low power radio—United States. 2. Community radio—United States. 3. Alternative radio broadcasting—United States. 4. Radio broadcasting—Political aspects—United States. 5. Radio broadcasting policy—United States—Citizen participation. 6. Pirate radio broadcasting—Political aspects—United States. I. Title.
HE8697.95.U6D86 2014
384.54′43—dc23
2014009827

Contents

Acknowledgments

Academic labor can at times seem lonely, irrelevant, or an exercise in being counted. But on most days, the opportunity to explore a topic deeply over time is an immense privilege, for which I am grateful. Foremost, I want to acknowledge the radio activists who are the subject of this book. They fearlessly and patiently allowed me to follow them around, asking questions and learning from them—for years. I have been deeply touched by their zeal and passion for their work, as well as by their openness to my research. Following ethnographic convention, I have not named individuals in the book, so I do not name them here either, but they should know who they are.

Later, after research was complete, I spent many hours on my own wrestling with prose and ideas, but in a very meaningful sense, I was never really alone; my debts for feedback, insight, and conversation are many. Ron Kline was a caring and thoughtful advisor for this project when it originated as a dissertation project. As unmistakably valuable as his guidance was during that period, I have in some ways come to appreciate his professional and personal support, as well as that of Trevor Pinch and Mike Lynch, even more in the intervening years since leaving graduate school; I am grateful for all they've done for me in school and beyond. Dominic Boyer also merits thanks for his involvement and encouragement during the genesis of the project and beyond.

Special thanks are due to Lucas Graves for always listening and offering suggestions that in countless ways have clarified and even uncovered my ideas for me, and to Laura Portwood-Stacer for patient readings at times when I was less so. Javier Lezaun also deserves special mention; my life and work would be vastly impoverished without his large and small provocations. Other comrades who have spanned a welcome continuum of friendship and intellectual companionship as readers and sounding boards during my years of work on this project include Chris Anderson, Peter Asaro, Biella

Coleman, Kate Coyer, Ingrid Erickson, Laura Forlano, Josh Greenberg, Dan Kreiss, Becky Lentz, Cyrus Mody, Rasmus Kleis Nielsen, Victor Pickard, Jeff Pooley, Christo Sims, and Todd Wolfson. Mentorship and kindness along the way were supplied by Michael Delli Carpini, Phil Napoli, Leslie Shade, and Tom Streeter. I am also grateful to my splendid colleagues and students at Rutgers for support and conversations about activism, scholarship, and other shared pursuits.

My dear friends Willy Schofield and Ian Lay, Vicky Karkov, Felix Teitelbaum, Chris J. Smith, Marc Washington, and Jesse Wilderman could always be counted on to bail me out mentally when I needed a break (more and less mentally—you know who you are!). I have benefited from the support of Ruth Rosenberg, Katherine Sharpe, Neerja Vasishta, and Steph Alarcón in the form of conversations about our various acts of production undertaken over these years; I hope I have supported them, too. Marshall W. Tidwell and Scott Edgar patiently tolerated harassment and kindly reciprocated when I needed distracting. As usual, Set Sokol didn't do anything readily summarizable, but I am glad for him anyway. Last but not least, Erik Hoversten has flitted around in the margins—prose and personal—for longer than anyone could have predicted. I thank him for many things including his indescribably enormous help editing this book in its final stages.

My two sisters, Anna and Sarah, often provide gracious counsel and indulge me in various ways, even though I am supposed to be the mature one. Elaine Dunbar and Al Hester, my parents, take interest and pride in my work and encouraged reading and questioning early and often. (My dad also encouraged listening to the radio, which seems to have had some effect, but not necessarily the intended one. My mom was more skeptical.) I appreciate my family's patience during my lapses in communication while I was finishing this book.

The National Science Foundation provided funding for my research travel (award 0432077), which was essential to this project. Any opinions, findings, and conclusions or recommendations expressed in this material are those of the author and do not necessarily reflect the views of the National Science Foundation.

I also wish to thank Margy Avery, Katie Persons, my reviewers for MIT Press, and the MIT staff for their enthusiasm about this project and their assistance in bringing the book into being.

Introduction

On October 4, 1998, a raucous group of protesters assembled in front of the Federal Communications Commission (FCC) building in Washington, DC. Seeking legal access to the airwaves for small-scale broadcasting by citizens and community groups, they engaged in established street theater tactics, including puppetry, chants, and speeches. In a less traditional move, they also flouted the regulators by broadcasting their protest into the building using a portable transmitter (it goes without saying, sans license). Of course this transmission was symbolic; the activists did not so much wish to instrumentally broadcast to the commission as to declare their presence on the airwaves and demand regulators' attention.

By 2000, their efforts had borne fruit. The FCC slowly began to issue licenses for new low-power FM (LPFM) stations; this was the first time in more than twenty years that would-be micro-broadcasters had a legal option for getting on the air. Several hundred new stations were broadcasting by the late 2000s. In the decade following the protest, a burgeoning movement for media democracy regarded LPFM licensing as a victory and mounted efforts along a number of other lines, including Internet governance, combating media consolidation, and securing support for public and independent media, to name only a few.

Yet low-power radio remained a primary concern for some. Many who had pressured the FCC to license "microradio" broadcasters continued to work to expand LPFM, albeit from a different position vis-à-vis the regulatory framework: with the possibility of legal broadcasting, efforts shifted to getting licenses into the hands of community groups, building new stations, and shoring up LPFM's status within telecommunications policy. The latter goal was attained in 2011, when President Obama signed into law the Local Community Radio Act of 2010, authorizing the FCC to grant licenses to additional new LPFM stations.

This book examines the practices of a small activist organization focused on LPFM during the early period of the institutionalization of LPFM, from approximately 2003 to 2007. The group had its origins in the mid-1990s as a pirate broadcasting outfit. But by the early 2000s, they had morphed into a non-profit organization to promote LPFM. The group engaged in a combination of advocacy to expand community media and hands-on technical work to build new stations (having ceased broadcasting themselves after being shut down by the FCC in 1998). This book traces their activities with an eye to the intersection of technical practice and political engagement. It specifically investigates how the radio activists imputed emancipatory politics to radio technology—notably, an "old" medium—against a shifting technical and political landscape that included increasing attention to Internet-based technologies. What is meant by "emancipatory politics"? Activists claimed that FM radio tinkering and broadcasting held the potential to empower everyday people through increasing democratic participation, autonomy, and self-determination at the community level. Their notion that expertise was accessible to all contrasted with more common conceptions of expertise; technology is more often constructed as the province of elite experts, and wider political, moral, and social issues are collapsed into seemingly narrow technical ones.

The politics of technology in media activism is a topic of more than academic interest. These radio activists are important because of their mediating position, situated between "upstream" regulators or policy makers and "downstream" user communities; they are not mere Luddites nor nostalgic hobbyists. Often, they attempted to exert influence in both directions, and their work to interpret, define, and propagate technologies has the potential to affect how ordinary users might understand, access, and make use of the technologies in question. Advocacy work to construct radio as highly local, noncommercial, and accessible to ordinary people had an impact on how policy shaped low-power radio. But the radio activists also exhibited a strong commitment to hands-on technical practice and work with radio hardware.

The sort of reflective technical engagement the radio activists promoted is significant for a variety of reasons. At the core of their technical practice was a commitment to a participatory politics, with attendant challenges and contradictions. In essence, though the radio activists claimed to favor radio as a medium for expression in part because of the ostensibly low barrier to access, attaching an emancipatory politics to tinkering and hands-on work was fraught. Though they valued technical practice as a means to demystify technology and create a political awakening in users, they struggled with the fact that patterns of inclusion and exclusion had already formed around electronics; historically practiced by elites, whites, and men, tinkering was not equally appealing to members of other groups. This tension between participatory ideals and expert forms of knowledge recurs throughout much of this book.

Anthropologist Jeffrey Juris, writing of anticorporate globalization movements, contends that "activists increasingly express their utopian imaginaries directly through concrete organizational and technological practice."[1] This is a useful starting point for understanding the practices of these radio activists, who were uniquely focused on technology and technical practice as the foundation for their vision of social change. In this book, I conceptualize the radio activists as "propagators" of technology.[2] I draw on the meaning of propagation as reproduction and replication and also its sense of creating an effect at a distance (and of course the entendre with radio wave propagation suits this group especially well).[3]

Although it is not unusual for activists to orient themselves around technologies as a part of a more extensive agenda for social change,[4] there are features that make these propagators unique. Propagators are special in how they combine mediation or interpretive work with a commitment to material engagement with an artifact. The radio activists hoped to place radio and their prescription for its use into as many hands as possible. Their goal was to set into motion social dynamics through the diffusion of radio technology and associated practices and then step back; they did not seek to oversee the these dynamics on an ongoing basis, instead believing that idealized social relations (including idealized media content) would flow from the act of propagation alone. Propagation was an act of knowledge production; in the radio activists' imagination, it produced not only hardware but also social relations. Propagation is thus articulating artifacts to politics and vice versa: while the radio activists were building technical artifacts, they were simultaneously building a politics of what might be called "participatory expertise." They strove to open up technical practice to people who were not technical experts. They understood this form of expertise to extend even beyond the domain of technology itself.

Put differently, activists turned to technology to express their political beliefs. At a typical technical workshop, people would spend hours in a basement soldering cables, then move to a rooftop to measure an RF (radio frequency) signal, before returning to the basement to try to fix a faulty connection or recalibrate equipment. One summer evening, radio activists and I moved from an electronics repair project in a basement to a scavenging project at a university engineering building slated for demolition. Having deemed it late enough to roam around the bowels of the building undisturbed, we spent hours digging through equipment that was being cast off. Cables, ammeters, and a horn antenna were among the haul. It was unclear what uses this gear would be put to, but its acquisition represented the values of reuse and repair, sharing, preparedness, and, of course, the requisite technical expertise to identify and imagine uses for the various pieces of equipment we uncovered. Reading a draft of this book, one activist commented to me that I "had written an anthropology of the basement." Her remark has a double meaning: the radio activists' office was literally in a church basement, a fact they made much of (and that served to distinguish them in their minds from more established nonprofit organizations). But she also marked the basement as a symbolic space of radio activism, which was not the halls or streets of Washington, DC, but the ubiquitous, grimy spaces of do-it-yourself (DIY) work and leisure. Radio activism was *everywhere,* and you didn't need more to participate in it than a soldering iron, your neighbors, and a basement. It was separate from but contiguous with everyday life, and accessible to everyday people. It challenged the separation of technical expertise from lay know-how, and technical practice itself was held to be transformative at the individual and societal levels.

Radio activists are not alone in tying their work with technology to politics. Internet governance geeks and free software developers can also be understood to engage in activism and deeply technical projects.[5] However, in spite of their similar technical commitments and normative claims, they largely differ from the radio activists: they usually achieve a consensus in which technical participation is limited to technical experts, which means they can focus more exclusively on debating and solving technical problems. They frequently leave the job of articulating the meaning of their technical work to mediating groups; mediators, rather than "techies," tend to translate technical projects and engage in advocacy. (The division of labor between Debian developers and the nonprofit organization Creative Commons within the free culture movement is one such example.) Propagators (who engage in technical practice and ongoing advocacy and mediation) are distinct.

Although a commitment to egalitarianism is not a criterion for the category of propagator, this commitment, however elusive in practice, further marks contrast between some other forms of activist technical projects and the radio activists. Plenty of activist projects around technology simply are not concerned with issues of unequal expertise. In many free and open source software projects, participation may be "open" in the sense that anyone who can contribute to a project is welcome to contribute. But a uniformly expert status among participants is unquestioned. By contrast, the radio activists were highly committed to drawing novices and laypeople into technical practice. However, they routinely found themselves confounded by the potential for conflict between engineers and laypeople, as well as by patterns of exclusion that ran against the egalitarian values they hoped to tie to technical practice.

Simultaneously, the radio activists were attuned to the fact that their project seemed anachronistic to some; their concentration on an "old," "dinosaur" technology seemed to belie their relative technological competence and sophistication. And yet "new media" were in many ways deeply and self-consciously implicated in the activists' propagation of radio. Radio activists—many of whom were well-versed in digital politics and activism—were concerned with alternatives to digital utopianism, resisting Internet-based communication as an analog (no pun intended) for what they understood to be salient and desirable about radio. This led to a situation in which they were, in some ways, defining radio in contrast to dominant ideas about digital media. They were especially interested in propagating an understanding of electronic media that emphasized local- or community-scale purposes, which stood in relief to the ostensible global reach of Internet-based technologies.

Though centered on radio, the dynamics this book explores are much broader. If we listen, articulation of values and political agendas to artifacts becomes audible. This radio case study is a model for other studies of technology. Too often claims about what the Internet is or does unquestioningly locate values and politics "inside" the artifact. Breathless exultations such as, "Digital technology can be a natural force drawing people into greater world harmony" (as stated by *Wired* magazine's Nicholas Negroponte in 1995) are ubiquitous across punditry.[6] The unbridled enthusiasm for "the digital" is not the only reason we should not accept these statements at face value. We need to recontextualize such declarations as part of a dynamic of articulation; they are rhetorical claims whose effect is to crystallize particular notions about what the Internet is. Radio activists' evangelism exemplifies how links are actively forged between politics and the technologies they engage. This phenomenon is as relevant to "the digital" as to older technologies such as radio. Indeed, it is only the hype-driven Internet mythology that causes us to think of anything associated with the Internet as new and anything analog-related (like radio) as old.[7]

During the period of my fieldwork, the radio activist group I followed faced organizational maturation and mission recalibration. They struggled to retain a sense of coherence in their work. Radical politics and technical engagement were of great symbolic importance, but these concerns did not in any way fully represent the range of projects the group was occupied with (which included everything from advocacy in Washington, DC, to researching health insurance policies for their organization, to fixing up old transmitters, to building radio stations). The radio activists pursued a unique combination of advocacy and technical work, which required a balancing act. Technical engagement nonetheless held a special symbolic value within their diverse repertoire.

Low Power to the People is organized as follows. Chapter 1 provides a historical introduction to my site, placing the radio activists into a wider cultural and historical context. Radio activism in this era must be understood as issuing from distinct yet interwoven social, cultural, technical, and political strands including: embedded practices of community media production and pirate radio; "Indymedia" and the transnational anticorporate globalization movement;[8] the emergence of "new media" including the Internet; and a regulatory environment favoring national broadcasting networks and corporate media consolidation that was opposed by a growing movement for media democracy.[9] Other antecedents to radio activism include ham and citizens band radio,[10] the Appropriate Technology movement of the 1960s and 1970s,[11] and earlier broadcast reform movements.[12] Even readers familiar with LPFM and its history will wish to read this chapter because it is here that I situate radio activism within earlier cultural formations and delineate the activists' specific priorities during my fieldwork. These priorities determined what I was able to observe and thus interpret: during this period, their attentions were split among fighting media consolidation, advocating for the expansion of LPFM, and building radio stations.

My own empirical work begins in chapter 2. This chapter takes up the issue of the activists' commitment to technical practice, focusing on their weekly tinkering group in their hometown of Philadelphia as well as the most significant symbolic site of their activism, the radio station "barn-raising." This was the name activists gave to their radio station-building events, in which volunteers and staff activists joined together to put a new LPFM station on the air over a weekend. Chapters 2 and 3 are devoted to the concept of "identity work," a thread that runs throughout much of the book (see also chapters 5 and 7). I show how the radio activists cultivated a geeky technical identity and how this interacted with other identities, including activist identity, countercultural identity, and gender identity. I argue that rather than existing as stable or inherent categories, these identities functioned as social tools; they were resources on which the activists drew, with varying consequences. The symbolic importance of technical practice for the activists also recurs throughout the book, especially in chapters 2, 3, 4, 5, and 7.

In chapter 4, I continue to explore the radio activists' technical undertakings, focusing on the productive and affective priorities of this form of technological activism. This chapter focuses on a pedagogical workshop the activists held over a weekend—the barnraising ethos in miniature, with attendant advantages and difficulties for activists, novices, and expert participants. I draw out some of the dynamics surrounding expertise that vexed the activists as they tried to realize a political vision that called for equality. I show that one consequence of promoting technology as a platform for emancipatory politics is that this can result in a constant battle with unequally distributed expertise.

I turn in chapter 5 to an examination of the role that technical affinity played as the group underwent organizational maturation. This largely played out as the systematic elevation of "technical" work and the downplaying of policy-advocacy expertise (even though both were salient in their work). I argue that the radio activists cultivated a technical identity that served to mark boundaries between their group and others in the terrain of media democracy work, which was especially important as they struggled to retain radical activist criticality and to resist being transformed into a "mainstream" nonprofit organization. At the same time, technical identity worked to downplay potentially troubling disjunctures within the activist organization. It marked continuity between the activists' past, present, and future, and it enabled them to assign coherence to a diverse range of tasks that might otherwise seem incongruent. I refer to this dynamic as "boundary effacement."[13]

Chapter 6 examines the discursive practices by which LPFM advocates attempted to redefine radio's use and meaning. During the 1990s and 2000s, radio broadcasting (a familiar and decades-old technology) remained the site of intense contestation (even in the wake of "new" media and Internet-based technologies). Echoing past reformers, radio activists (and other advocates with whom they were not always in full accord) defined FM radio as noncommercial, well-suited to local or community-level use, and a medium for political expression and organizing. The activists' role as propagators of technology was evident as they sought not only to diffuse the artifact of LPFM, but also to shape interpretations of broadcasting.

The final empirical chapter of this book follows radio activists' assessments of emerging Internet-based technologies (primarily wi-fi networks). Particularly for urban areas where LPFM licenses were out of reach, the activists considered other "appropriate technologies" to promote citizen media production. I show that the radio activists were selective in their adoption of or resistance to various options, some of which they largely rejected (such as webcasting) and others of which they cautiously embraced (such as community wi-fi networks). Having identified radio as the artifact with which their politics best aligned, they were circumspect about the propagation of other technologies that were less obviously tied to the values they identified in radio.

In the book's conclusion, I argue that in their efforts to define and propagate radio, the activists demonstrated an understanding of media technologies as tools for transmission of information, promotion of "community," and redistribution of power. But I contend that they privileged promotion of community and redistribution of power over transmission of information. This stands in contrast to some strains of digital utopianist thought, which view access to electronic communication technologies as tools to disseminate information. Here the value of analyzing contestations around an old technology becomes more apparent: listening to the debates around contemporary meanings of broadcasting helps us tease apart differing values surrounding electronic communication in general. This book's final point is that because technical expertise is unequally distributed, there are real risks in fetishizing technology as a platform for egalitarian politics.

What This Book Does Not Do

First and foremost, this is not a book about the future of radio, the end of radio, or the future of digital technology. In some ways, it is not even a book about radio; rather, it is a book about questions that precede all of those topics: how do certain artifacts come to have particular political meanings? Do certain beliefs about the role of technology in human affairs lead to particular choices about technology? Do politics have artifacts?[14] And, in which ways do people's close relationships with technologies (including tinkering and pleasures in technology) come to shape wider interpretations of technology?

This is a book about people tying politics to artifact—it examines the construction and implementation of specific beliefs about what technology can do, what technology should do, or what artifact is most appropriate to enact a set of politics. This story offers wider lessons for scholars of technology and broadens the relevance of this topic beyond the issue of radio per se. That said, the fact that radio activism is the object of study is of course wholly relevant: this case is not presented as a generalizable one, but it is an especially interesting one in part because it affords an opportunity to trace the trajectory of a familiar technology into its ninth decade of existence (roughly locating the origin of broadcasting, as distinct from radio telegraphy, circa 1920). Far from being a "settled" technology, radio reverberated with renewed vibrancy and relevance in the contestations that I chronicle.

The media institutions and artifacts we have (as well as those we have had in the past and might have in the future) are in part the product of political activities oriented to certain policy goals. Although policy ideas matter very much in the shaping of media technologies and institutions, this is not a book primarily about policy. Though telecommunications policy was of course hugely implicated in what the radio activists were enabled to do (or how they were constrained), here I am less interested in commenting on policy or critiquing dominant policy discourses. Instead I take an approach consistent with what anthropologist Hugh Gusterson has called "a cultural perspective on a policy problem."[15]

I am concerned with how people structure beliefs around artifacts (or how politics form around normative ideas about what technology *should* do). I ask how and why these activists sought to produce the social and technical arrangements they deemed most desirable. My aim is to explore the lifeworld of low-power radio activism, and in so doing shed light on the cultural processes involved in activists attaching a particular emancipatory politics to a decades-old communication technology. I show the radio activists' efforts to construct meanings for radio broadcasting that include its viability for local communities, its potential as a political medium, and its continuing vibrancy at the turn of the twenty-first century. I seek to explain the significance of their pursuit of radio in an era when many dominant discourses had become focused exclusively on "new media" and "the digital."

It also must be stated at the outset that though my object of study—technologically oriented media activism—is related to what might be termed a social movement, my analysis is not at the level of a movement. Rather, I examine media activism at the level of practice, seeking to understand how social groups attempt to build politics around communication technologies and vice versa. I ethnographically interrogate activism as a creative and productive activity, even while acknowledging that actors' practice was often related to their self-understanding as being members of a movement. To refer to this movement, I use "media democracy movement" most frequently, but it is important to note that this is a disputed label. Indeed, the issue of whether this is actually "a" single movement is also contestable.[16]

I am concerned with generating a practice-oriented understanding of activism, as opposed to taking activism as a category for granted. Though Edward Woodhouse et al. usefully define "activism" as "a range of methods used by groups with relatively little institutional power attempting to influence opinion, policy, or practice,"[17] this sheds relatively little light on what the granular practices of activism *are*. In particular, how may some activities that look very different from one another all be understood to be activism? Or conversely, how may activities that appear similar from the outside be distinguished as activism—or not activism—according to actors' understandings? This work also differs from other studies of media activism that offer meso- and macro-level analyses addressing typologies of media activism, as opposed to practice.

Scholars have pointed out that there are some key differences between "conventional activism" and media activism, because conventional movements seek to use the media instrumentally in pursuit of their agendas, whereas media activists view media as an end in itself.[18] The media democracy movement is thus an "umbrella" movement: people working around other issues often come to media activism after identifying media as a linchpin that will constrain or enable their organizing around their original topic. This movement's constituency is wide and not always exclusively committed to media issues. Much of the literature on new social movements has tended to explore the construction of collective identity.[19] William Carroll and Robert Hackett argue that media activism in particular lacks a "clear, regularized collective identity,"[20] in part because of the way it serves as an umbrella or perhaps meta-movement. I do in some way share with theorists of social movements an interest in identity. But rather than attending to how people construct and negotiate collective identities within a movement, I focus on how differing commitments and identities interact with technical practice (particularly at the intersection of political agency and technological engagement). I examine the local practices of activist work and identity construction, without much regard to how these might play out at a movement level. Though these iterations of identity (and the complements and conflicts between them I uncover) might very well be relevant at a movement level as well, those dynamics are not a focus of this book. My interest in affective relationships people form with technologies may have resonance with cultural approaches to social movements that seek to understand the construction and role of affect in social movements, but is not directly analogous.[21]

Research Activities, Methods, Position

This project combines a deep, single-site ethnographic inquiry focused on one group of radio activists with interviewing and observation at other critical sites. My fieldwork began in 2003. I spent that summer in Philadelphia conducting interviews and participant-observation with a group of media activists who gathered weekly to tinker (referred to as "Geek Group"). I identified Prometheus Radio Project, a small organization devoted to radio activism, as my main field site during this period. Prometheus stood out among media advocacy groups (others of which do not generally combine policy advocacy with hands-on technical work). They provided an excellent focal point to engage themes related to politics of technology, technological engagement and skill, and negotiation between technological options. I volunteered with Prometheus in 2004–2005 and immersed myself in participant observation, which included working and observing in the office and observing activities in Philadelphia. I also accompanied Prometheus on trips to Washington, DC, to attend meetings with lobbyists, FCC members, members of Congress and their staffers, and community groups seeking to obtain LPFMs. We also traveled to Chicago multiple times to meet with community members and advocacy groups working on wi-fi networks. I spent a month with an organizer on a speaking tour across parts of the Midwest and the South. I attended LPFM radio station barnraising events in Pasquo, Tennessee and Florence, Massachusetts, as well as a similar station-building event in a village outside of Arusha, Tanzania, in East Africa. In 2005–2006 I pulled away from full-time participant observation, but I conducted additional interviews and I continued to attend and observe special or significant events, including one additional barnraising in Woodburn, Oregon, in 2006. I also attended Prometheus workshops on community wi-fi and proceedings on municipal wi-fi in front of the Philadelphia City Council. In all, I conducted twenty-nine semistructured interviews, most of which were recorded digitally and manually (I took detailed notes that I used to guide me when I went back over the audio recordings); I conducted additional informal interviewing in settings such as barnraisings. These interviews included not only activists but also members of regulatory bodies (the Federal Communications Commission and Congress), the corporate broadcast lobby, and other media reform advocacy groups. I also met with LPFM radio station station holders and volunteers, academic institutions, and foundations that fund media reform work. I complemented my ethnographic fieldwork with documentary research on activism and policy from 1996 to 2006. This included comments and petitions filed at the FCC and documents produced by advocacy groups, National Public Radio, and the telecommunications trade association, the National Association of Broadcasters, all of whom weighed in on the shaping of LPFM. This decade, marked by the Telecommunications Act of 1996 and the decision of the Federal Communications Commission to start licensing low-power stations again, was crucial to the evolution of media activism in the United States.

According to Wiebe Bijker, "Actors provide an effective starting point from which to identify relevant social groups"[22] for understanding how technologies acquire specific meanings and importance. My account follows the activists' involvement in contestations around radio but does not attempt to "balance" their concerns with those of other social groups, including regulators, media policy groups, incumbent broadcasters, lawmakers, members of the public, and groups who desired to broadcast. Instead, I take the relational positioning work on the part of the activists as they attempted to influence other groups as central to the analysis.[23] I allow the activities of the radio activists to largely define my priorities in terms of identifying other groups relevant for inclusion in this account. Other groups' voices can be heard most plainly in chapters 5 through 7.

Ethnographic truths are always partial. Though I observe and make an effort to narrate the events in this account with a commitment to fidelity, my account is inherently incomplete.[24] (Here my invocation of "fidelity" evades simplistic realism; although I build this account on carefully conducted research, I do not seek to "merely represent" the activists' milieu, nor do I consider this "objective" mode of representation to be possible or desirable.) I am aware that my presence in research settings had the potential to actively change what I was studying.[25] Activists would on occasion question me about what I was taking notes on. They would occasionally actively point out "important" matters to me or comment that a particular setting or event would be "especially interesting" for me as an analyst; other, more subtle instances also doubtless occurred. All of these factors, as well as countless others, inform the analysis I am able to present. As anthropologist Hugh Gusterson writes, echoing Donna Haraway, "there is knowledge here, but it is ... situated knowledge."[26] I do not claim to offer a complete or generalizable account. I am instead presenting ethnographic "true fiction."[27]

Anthropologist of science Sharon Traweek writes, "The fieldworker needs to remain marginal. If she were to become a fully integrated participant in the community, its sociocultural assumptions would no longer stand out in the foreground of her attention; and in any case it would no longer then be appropriate for her to be asking questions about the meaning of social actions."[28] I approached researching the radio activists as a novice to electronics and with no skills in journalism or audio production. My greatest exposure to radio production prior to this fieldwork was occasionally sitting in on the monthly radio show my best friend had in college. I did, however, have office and writing experience and a distant background in youth activism (though nothing that resembled the organizing campaigns at Prometheus). In some ways, it seemed a hindrance that I was not more versed in the skills of the group, because I could then perhaps have contributed more fully to the projects and work undertaken by the activists. But at the same time, my relative unfamiliarity was a benefit in terms of being able to make critical sense of social actions, as mentioned by Traweek. Because many of the group's events (including barnraisings and Geek Group) were purportedly about imparting skills, there was merit in learning as an active participant, rather than trying to reconstruct learning through interviews or observation. There would also be less pedagogy to observe if everyone in a group were relatively expert, so novice status was additionally useful in that regard. In terms of the analysis I make on gender (as well as race and class, though gender receives greater attention here), my own positionality as a middle-class white woman pursuing an advanced degree is certainly worth consideration. I cannot say that my experience in the group was a universal one, not least because I was there as an observer and social scientist. I was also a female person trying to learn. The geeks' treatment of me, and interest in my experience as a novice, woman, ethnographer, and participant in the group of course contribute to the observations and analysis I offer here.

Especially at barnraisings, it was often difficult for me to stick to my own research agenda. I participated as "ethnographer" but also as a volunteer, not least because of the fact that so much work needed to be accomplished. Through mutual negotiation, the activists saw me as somewhat beholden to them for ethnographic access. There are multiple ways to characterize this dynamic: because the dominant feature of this environment was purposive (and often frantic) activity, not being on call to pitch in would have been puzzling to the activists, if not outright offensive. I was often concerned about appearing diligent and engaged, even if I did not feel self-directed and un-self-conscious in my activities. I felt strongly that being too self-directed in supporting the activists' mission was contrary to my agenda as a researcher. Even though I became more comfortable with the group over time, I was often internally conflicted and anxious about these dynamics. Nonetheless, at their request, I ran errands, moved furniture, cleaned, staffed registration, soldered cables, hammered nails, and once, helped run ongoing soldering work at a barnraising (more responsibility than I desired or intended to take on). I also dropped into workshops and work stations to observe, ask questions, and help out.

In many social studies of science and technology, "fieldworkers enter scientific fields which they do not know, and try to learn enough about them to do sociological analyses. Rarely, however, do they reach the level of expertise of a full-blown participant. In the case of the esoteric sciences, the fieldworker hardly ever participates in the science itself."[29] Given that the radio activists held a decidedly different attitude toward expertise than do members of the "esoteric sciences," as I learned more, they were comfortable putting me in leadership roles in tasks I was not even sure I was qualified to do, especially when they put me in charge of the soldering track at a barnraising. This gave me a valuable firsthand glimpse into some of the activists' notions about leveling expertise. I admit to feeling some discomfort at being put on the spot when more expert "geeks" would look to me first to answer questions posed by newcomers about transmitter components. At the same time, there were often times when I felt that the activists' perception that I was able to contribute to "productive" activities detracted from my autonomy as a researcher. At the barnraising where I was teaching others to solder, I was not as free to circulate with my notebook. I felt my agenda as a researcher strained by having been enrolled into the pedagogical dynamic I was ostensibly studying. At the next barnraising I attended, I deliberately hung back more and tried to remain more autonomous. I did not commit to any single work activity that would prevent me from moving around freely. I doubt my malingering was noticed in the crush of activity, and in any event, I felt (perhaps unjustifiably) that I had "earned" this liberty after having permitted the activists to define my agenda at the previous barnraising.

In any event, maintaining some distance was of value. Another choice about "distance" had to do with my deliberate decision to distance myself and therefore my analysis from the occasional workplace squabbling that arose at Prometheus. As is likely common in organizations of this size, the members struggled with organizational growth and defining appropriate priorities and divisions of labor. Only one full-time person had been with Prometheus since the beginning, and he had something of a "charismatic leader" quality to him. This manifestation of so-called "founder's syndrome" presented some difficulty for newer staff organizers, whose contributions to the organization's work and trajectory were also significant. (It was at times troubling for this staff member as well, as he sought to delegate responsibility and support efforts to make the organization self-sustaining and less dependent on the vision and legacy of a single individual.) Hoping to avoid being drawn into these conflicts, I found that occasional offers of reassurance that I was not intending to air Prometheus's "dirty laundry" soothed a newer organizer in particular. Not only did this smooth our interactions, it felt like the least I could do, given that the activists were generally incredibly open, tolerant, and trusting toward me. This was the case even when I asked them to share their experiences with potentially sensitive subject matters or to critique their own actions or campaigns. I do not feel that this relative remove hindered my ability to conduct a project that assessed the intersection between technological negotiation and political agency. The decision to remain essentially willfully ignorant of the details of interpersonal flare-ups may have helped me analytically. I did not feel that I was ever close enough to any particular individual to be perceived as being on any particular "side" of any particular conflict.

We might characterize my experience in the field as one in which I attained interactional expertise, sufficient "to interact interestingly with participants and carry out a sociological analysis."[30] Yet unlike studies in which the researcher "studies up," studying actors with more social capital and status than the researcher, I might characterize my experience studying activists as "studying sideways," a concept I borrow from anthropologist Ulf Hannerz. Interestingly, when I began the study, one of my key informants and I had the opportunity to converse about topics familiar to science and technology studies (STS), because he had, in a self-designed undergraduate major, been exposed to canonical works, including those by Bruno Latour and Donna Haraway. I wonder if his familiarity with this scholarship made him more open and at ease with me. Certainly his comfort with me paved the way for my acceptance by the other activists. As a doctoral student, my similar age, economic status, and level of educational attainment placed me on similar footing to the actors in many ways. Social scientists and activists (similar to journalists, the objects of Hannerz's research) are engaged in social mediation and cultural production.[31] Throughout my research, these juxtapositions generated absorbing conversations; I unexpectedly ran into a radio activist at an annual meeting of the Society for Social Studies of Science that she was "checking out"(!). She later remarked to me that "[if] technology is the Trojan horse for social and political agendas, we [activists] can play that game too, position ourselves as technology experts."[32]

Nonetheless, my status as an "academic" still marked me as different. I perceived this in many ways, including having another informant regularly bring up my practice of note-taking, feeling differences in my own personal and professional style of work from that of the activists (which, had I not reminded myself that I was there to observe, I might have found more frustrating), and being occasionally introduced as "our anthropologist." (Even though I did not identify myself as having a particular disciplinary affiliation beyond my doctoral work in STS, activists occasionally would playfully refer to the popular understanding of the anthropological practice of "studying tribes" as a way to make my presence comprehensible. One activist expressed this attitude when he wrote in an e-mail, "It is great to be part of the onward march of science, if only as a lab rat!"[33])

This lighthearted marking of "difference" between activist position and academic position belies potentially profound tensions. As noted previously, activists and scholars are united in social and cultural mediation. Yet their goals, methods, and products often vary intensely. In some ways this is a subset of a tension common in ethnography; a researcher will often embed herself deeply enough in the practices of the community to gain a feeling for its priorities, practices, and values. But fieldwork, even if not fully bounded or discrete, ends. And indeed, in the moments at which the researcher's focus shifts to reflection, narration, and abstraction, her priorities are brought into contrast and even conflict with the material and political commitments of actors she in these moments represents.

This is not to suggest that academic projects always differ from activist ones. Academic research may overlap with activism, or be conducted in sympathy or solidarity. Yet the difficulties collaborating across these communities of practice are real. In the realm of media democracy, I have heard people in nonacademic roles raise the difficulty of understanding academic writing. Some questioned whether engagement of academics with activist-advocacy topics was conducted in sympathy with them, because it seemed impossible to determine this from the academic products. In this project, I was fortunate to be received by informants who were sufficiently open and trusting toward me to allow us to have productive, reflective conversations and to avoid conflict. I arrived at this project with political commitments that in some ways mirror theirs: I care greatly about the struggle for a more just and more equal social world. I have a strong sense that the current US media system, built to support and extend corporate interests, is a missed opportunity for social change, and thus is ripe for meaningful intervention.

Nonetheless, my interests ultimately lie not with "the media system" but with people's relationships to technology. My scholarship centers on how these relationships are inscribed in our social and material world. As Bruno Latour writes, "technology is society made durable." And thus my interests tend to be more analytical than instrumentally oriented. This is not an easy position to occupy—I am in fact drawn to study activism because of an affinity for many activist goals. Sometimes I wish I could more embody a deeply proactive stance (beyond that of critique). Yet by temperament, I ask questions and reflect. This is a way of saying "must it be so?"—an act of questioning and ideation that also drives activism. But mine is a different route and produces different outcomes. The radio activists are better at asking (and answering) "must it be so?" about the media system (and society) into which they hope to intervene than I could ever hope to be. Nonetheless, I hope that what I may contribute is a different layer of cultural mediation, one that is not incommensurate with the activism I interpret. What I offer is a reflection and critique of some activist methods; in particular I interrogate the consequences of tethering a politics of empowerment to technical practice. I do not wish to reduce activist practice to a "spectacle,"[34] but my terms and commitments are not identical, either. My hope is that there are some lessons in my accounting for those who would identify technology as a platform for social change. I also hope that my critiques and activists' technical interventions can be juxtaposed in productive and collaborative ways to advance commitments we share.

On Names

In this book, I refer to the primary organization by their actual name: Prometheus Radio Project. I use pseudonyms to refer to individual actors associated with the organization. This represents a shift from earlier presentations of this work, where the organization was thinly pseudonymized.[35] My goal has never been to deeply anonymize the group, but rather to separate people's identities as individuals from the comments and actions I narrate and interpret.[36] It would be close to impossible to fully anonymize the group because they are too individualistic; no one else combines policy and technical work on low-power radio the way they do. The group's actual identity would doubtless have been discerned by readers familiar with US media activism anyway. (Even their participation in pirate radio is part of activists' public narrative. A Human Subjects Board review rightly flagged the issue of potential identification of individuals involved in illegal activity. But in this case, the radio activists were already "out" publicly about this activity; their pirate history routinely surfaced in presentations they gave and in media accounts.) In naming the organization but not the individuals, I hope to provide fidelity to the historical record. Prometheus is widely known in media policy and activism circles, as well as being the plaintiff in the lawsuit *Prometheus Radio Project v. FCC*. As a result, it seemed nearly absurd to strike their name from my account. At the same time, I wanted to adhere to the longstanding ethnographic convention of not naming individuals. Pseudonyms provide a layer of artifice that protects individual people from having their true identities associated with their every utterance, decision, and action.

Whether or not to name the group here was not a straightforward matter. In a true sense, the site that I studied no longer exists: between the period of my fieldwork and the writing of this book, the activists' primary legislative goal during the 2000s was at last accomplished. The political climate in which they were working changed with the election of a Democratic president. The organization further matured, and staff turned over nearly completely (to say nothing of volunteers and interns, who constantly paraded in and out). In some ways the period represented by this book now seems like an alternate reality that is of interest because of the priorities sealed within it. I am relatively unconcerned with connecting these events to the present or predicting the future. Instead I offer a narrative about people tying politics to an artifact based on specific beliefs about what that technology can and should do, situated within and flowing from their understanding of a particular historical moment.

Because the events I recount can now only be partially reconstructed, in ways that reflect my own priorities, I present the actors as players in an ethnographic true fiction. The primary informants are always pseudonymous, but some peripheral organizations and people with public profiles who were not primary informants are called by their real names. The distinctions are not always entirely clear-cut. This acknowledges that although real events that occurred are at the core of my analysis, I have often blurred the sites and especially the players just a bit. But as the focal range shifts, my more micro story bleeds into matters of historical, journalistic, and policy record (and thus extends beyond the borders of my more intimate accounting). Disguising the features of the wider social movement and policy issues would render radio activism unintelligible, so I do not attempt to do so. One main informant said of an early draft of this ethnography that it was a struggle for him to read through because it was "like reading your therapist's notes from five years ago." This reaction alone seems like another reason to offer a veneer of anonymity. An interviewee commented that the draft had an "allegorical aura about it," specifically relating to how I employed anonymity. I took his reaction as a signal that I had struck an appropriate balance in disguising the most particular aspects of this story while retaining the broad meanings of events and interactions I narrate and interpret.

1 Pirates, Hams, and Protest: Radio Activism in Historical Context

One year after the mass actions in protest against the World Trade Organization's meetings in Seattle in 1999, Philadelphia braced to host the Republican National Convention (RNC) in August 2000. In anticipation, activists formed an independent media center on an ad hoc basis. Independent media centers (IMCs) were rhizomatic[1] citizen-journalist media centers devoted to creating and disseminating alternative news content within a network known as "Indymedia."[2] They were founded in reaction to neoliberal ideology and globalization. IMCs sprang up all over the United States, and indeed the world, at the turn of the twenty-first century. The earliest were founded in the late 1990s at the time of the Seattle protests. One estimate was that there were about sixty in the United States in 2005.[3]

In Philadelphia, the IMC formed in response to a perceived need to counterbalance mainstream media coverage of the RNC. In particular, it focused on legal protest activities, which were largely ignored by mainstream coverage or covered in ways that were unsatisfactory to activists. The activists claimed that a local television station had told them that "it is against [the station's] policy to cover 'staged political events'"—willfully ignoring the irony that the RNC is a staged political event of the highest order. This anecdote may or may not be apocryphal, but it points to dissatisfaction with mainstream coverage and the perceived need for alternative media's "counter-hegemonic textual products"[4] as exemplified in IMC coverage. In order to cover the RNC, the IMC established a bank of volunteer reporters and set up a website for print, audio, and video content. They also launched an unlicensed FM radio station to broadcast for that week. Many people found this experience to be a powerful and galvanizing one. They committed to making independent media activities a more stable and permanent part of the activist landscape in Philadelphia after the convention was over. It was at the events around the RNC that I took note of the widespread activist interest in media technologies. In particular, their claims that equated access to technologies with democratic participation, community autonomy, and self-determination were of obvious analytical interest.

A wider movement for media democracy in the United States was spurred by the Telecommunications Act of 1996, which permitted unprecedented consolidation of media companies. Greg Ruggiero states that "the Act's defining feature [was] the toleration of a higher limit of media outlets—radio and TV stations—that any corporation [could] own. It also ease[d] restrictions preventing these huge media conglomerates from merging into one another...."[5] For radio, this was particularly evident, because of the relaxed prohibitions on ownership of radio stations. Whereas the pre-1996 national limit was forty stations, by the early twenty-first century, one single company, Clear Channel Communications, had acquired more than twelve hundred radio stations—a *3000 percent* increase in only five years.[6] Even prior to the 1996 act, citizen[7] scrutiny of radio regulation was already intense. The FCC had ceased to grant low-watt noncommercial licenses in 1978, to the dissatisfaction of many who desired to broadcast. Activist pressure on the FCC mounted in the 1980s and 1990s. Unlicensed, illegal broadcasting proliferated, and court battles between the FCC and unlicensed broadcasters ensued. By the 1990s, many of these participants viewed themselves as part of a "micro-broadcasting movement," devoted to pressuring the FCC to grant low-power noncommercial licenses. In some cases, these activists challenged the right of the government to license the airwaves in the first place.[8]

The intersection of activism, media technology, and protest that occurred in the year 2000 was not a new phenomenon in Philadelphia. Some who participated in the RNC convergence had been part of a "pirate" broadcasting collective, Radio Mutiny, in the 1990s.[9] Radio Mutiny was raided and shut down by the FCC in 1998. After the Mutineers' station was terminated, members went on to form an activist nonprofit group called Prometheus Radio Project. Prometheus sought to provide technical and legal assistance to groups wishing to apply for licenses and to set up legal LPFM stations. Prometheus also closely followed regulatory activities in Washington, DC, and advocated for LPFM. A founding member was a participant in drafting the rules that initiated the LPFM service in 2000. This book follows the activities of Prometheus and others with whom they interacted, based on fieldwork mainly conducted from 2003 to 2006. Thus, the context for the activities that are the focus of this book includes 1990s micro-broadcasting, Indymedia, and the more general media democracy movement that formed in the wake of the 1996 act.[10]

Media activism is sometimes viewed as an end itself, but often people interested in media activism are first involved in other social justice issues. They identify media access as a key component of activist work on any issue; "media activist groups tend not to respect existing [social] movement boundaries, but to exceed them."[11] This was the case for the Philadelphia group, whose members were active in various causes, including housing rights and ACT UP (the AIDS activism group), before concluding that their efforts were essentially futile without a media system where they might circulate their causes and ideas. One person stated,

> The whole reason that ... I lean towards media democracy movements and struggles is that when I moved out here [Philadelphia], there were so many causes I wanted to be involved in, and I never could have done all of them, I felt like I was flooded with requests for help....
>
> A big problem [for] a lot of activists is that the more you get involved, the more you see how fucked up everything is, and how you really have to change everything in order to change one thing.... I thought that building [a radio station so] they could have their own show[s] would be a way to help everybody that I wanted to without focusing on one thing.[12]

This is a fairly representative viewpoint among people whose goal is media change. At the 2005 National Conference for Media Reform in St. Louis, Missouri, plenary speaker Malkia Cyril, director of Oakland-based Youth Media Council, echoed this sentiment: "For people of color, queer people, women, and young people, there has never been a free press, and without racial, gender, and economic justice there never will be." She led the audience in a chant ("When I say 'Media!' you say 'Justice!'") to illustrate the belief that these issues are deeply intertwined.[13]

This should not be taken as an indication that only people with leftist politics are concerned about media consolidation. In fact, the groundswell of opposition to then-FCC chairman Michael Powell's June 2003 recommendations to allow further consolidation united people and groups across the political spectrum. This made for such strange bedfellows as Prometheus, the National Organization for Women, and the National Rifle Association, all of whom mobilized their constituents to oppose consolidation. Groups that opposed media consolidation did so for many reasons. Many groups saw consolidated commercial media as standing in the way of other uses of a media system, including cultural and musical programming free of corporate control and community-based news and public affairs programming. Advocates understood broad and direct citizen access to media production as a way to a support meaningful and vibrant public sphere.

But why radio? Some view radio as unique among media technologies: it does not require producers or listeners to be literate; it can reach a small, local community or area; production and broadcast technologies are relatively inexpensive and easy to use; radio is very inexpensive to receive; and it is easier and cheaper to provide programming in an aural-only medium than in a televisual one. In spite of charges of that radio is a dead or dying medium—an allegation as old as the advent of television[14]—policy advocates, radio activists, and corporate broadcasters view the FM band as valuable, even into the twenty-first century.

In fact, the commercial broadcast lobby and National Public Radio (NPR) opposed the introduction of the LPFM service in the late 1990s and early 2000s. They argued that the issuing of new licenses would overcrowd the airwaves and interfere with the service and transmissions of incumbent broadcasters. Many LPFM advocates saw this "technical" concern as motivated by a political agenda or fear of competition. The new LPFM service introduced in 2000 was almost immediately decimated when Congress, at the behest of the broadcast lobby, held up issuing many licenses in order to study the issue of interference. Ten years later, the original LPFM service had not been restored, though about 850 new LPFM stations were on the air by the end of 2010.[15] In early 2011, President Obama signed into law the Local Community Radio Act of 2010, which would permit additional new LPFMs to be built.

Low-power FM (LPFM) is a legal designation in the United States, used to describe FM transmitters that operate using between 10 and 100 watts. This is enough power to reach a few square miles from the site of transmission at best. *Low-power radio, free radio, community radio,* and *microradio* are all terms that refer to noncommercial radio usually broadcast at a low wattage (though not all community radio stations are low power). I use "LPFM" or "community radio" to indicate legal stations, and "unlicensed" or "pirate" to specify broadcasting that is illegal because of its unlicensed status.[16] The label "low-power FM" signifies that the radio station is legal and was created after 2000. LPFM specifically refers to locally owned, noncommercial radio stations that tend to broadcast a significant amount of original, locally produced content.[17] Microradio and free radio stations are technically similar to LPFMs in terms of their wattage and range, but they are usually unlicensed,[18] and preceded or coexist with the LPFM designation. Both terms connote 1980s and 1990s unlicensed broadcasting intended to challenge the FCC. Unlike micro- or free radio participants, "pirate" operators are less easily categorized as similar to community radio or LPFM. Some pirates share with free radio and community radio the ideals of democratized communication and community access, but others may broadcast for any number of reasons. If we listen to LPFM, it becomes possible to hear reverberations of older contestations about the use and meaning of small-scale radio and even radio broadcasting more generally. That being said, LPFM as a particular historical artifact has existed only since 2000.

Regulatory History: A Brief Review

By the 1920s, radio had been defined as a medium for broadcasting; by the 1930s, it was largely networked and commercial. The FCC created the noncommercial educational FM band in 1938 in order to comply with the Communications Act of 1934, which ordered the FCC to allocate spectrum for noncommercial use.[19] FM had some technical advantages over AM, such as the potential for high-fidelity stereo sound and less static and interference.[20] After World War II, the FCC shifted the FM portion of the spectrum (from 1 to 42 mHz to 88 to 106 mHz), rendering the pre-1946 FM system obsolete (the 1 to 50 mHz band was reallocated to fixed, mobile, and land communication). This contributed to the slower adoption of FM. AM's stability during this period gave it an advantage.[21] The FCC first began issuing licenses for what were called "Class D" (10 watts or less) FM stations to educational institutions in 1948, hoping that this would encourage educational institutions to populate the largely vacant educational FM band.[22] The 1950s actually saw a decrease in commercial FM stations;[23] by the mid-1950s, the FM band was largely used for duplication of AM radio signals, and was not considered commercially valuable.[24] In contrast to commercial FM stations, the number of noncommercial FM stations on the air, including the 10-watt Class D stations, grew steadily beginning in the late 1940s, with 311 noncommercial educational stations on the air by the late 1960s, 134 of which were class D.[25] In the mid-1960s, the FCC ruled that companies owning stations broadcasting over both AM and FM had to program AM and FM separately. This forced the FM band into a period of reconsideration and experimentation, enabling its technical properties (such as the potential for stereo sound) to be explored and exploited. Also in the mid-1960s, the consumer demand for hi-fi stereo equipment grew.[26] As a result, the FM band came to be perceived as more valuable commercially, steadily expanding throughout the 1970s. In 1979 FM's listening audience exceeded AM's for the first time.[27]

Multiple broadcast reform movements accompanied these changes in radio technology and policy. Two unsuccessful reform movements (one in the early days of broadcasting and another after World War II) preceded a successful one that began in the 1960s.[28] This third movement was broadly linked to the civil rights movement and included a landmark lawsuit in which the United Church of Christ sued the FCC over the issue of license renewal of a segregationist television station in the Deep South.[29] President Lyndon Johnson signed the Public Broadcasting Act of 1967 as part of his Great Society reforms. This act created the Corporation for Public Broadcasting (CPB) and stated that "it is in the public interest to encourage the growth and development of public radio and television broadcasting, including the use of such media for instructional, educational, and cultural purposes."[30] The impetus for this act was in large part disenchantment with commercial television. It aimed to create alternatives to the existing commercial television system, which were imagined to be possible with the advent of cable television and the increasing accessibility and portability of video equipment.[31] The act also affected radio. In 1970, the CPB established NPR, a nonprofit membership organization comprised of public radio stations. NPR absorbed the National Educational Radio Network, a precursor association that included college stations.

During the 1970s, the CPB and the FCC sought to meet the growing demand for radio stations in the FM band. The CPB saw the 10-watt Class D stations as "threatening to exhaust available frequencies, preventing high-powered 'full-service' stations from getting on the air."[32] It urged the FCC to address this issue. Though nonmember stations argued to retain their independent, small-scale status, they ultimately were less persuasive to the FCC than NPR and the CPB, who sought to consolidate the FM band (in large part to expand NPR through absorbing existing stations into the network). In 1978, the FCC passed an order requiring 10-watt stations to increase their operating power to 100 watts or be subject to encroachment (rendering them without protection from the FCC if a full-power station's signal interfered with theirs or wanted to move to a frequency that would interfere with their signal). The commission additionally stipulated that no new 10-watt stations could be created.[33] Thus from 1978 onward, the FCC ceased to offer noncommercial small-scale licenses for the sorts of groups that had used them from 1948 to 1978, such as educational institutions and community groups. Access to FM licenses was limited, and the FM market came to be perceived as increasingly valuable.[34]

In the 1980s, and increasingly in the 1990s, some people who objected to the FCC's refusal to grant licenses to community groups undertook unlicensed broadcasting in acts of "electronic civil disobedience." Forming a loose movement, they helped one another go on the air, even as the FCC threatened to (and did) shut some of them down. The issue also received attention due to some well-publicized lawsuits (notably Mbanna Kantako's skirmishes with the FCC beginning in the 1980s, and Stephen Dunifer's Free Radio Berkeley 1990s battle with the FCC in the California courts). Although it is obviously difficult to get an official count of the number of unlicensed broadcasters, some estimated that there were one thousand in 1997.[35]

By the late 1990s, the FCC recognized this growing movement and admitted that these "pirates" might have a point. Then-FCC chairman William Kennard stated that he was "'receptive to hearing' about models for legal microbroadcasting."[36] In January 2000, Kennard and the FCC initiated the LPFM service, creating the opportunity for community groups to get noncommercial FM licenses to broadcast using from 10 to 100 watts (a few miles). Since then, the issue has remained disputed. In December 2000, the newly created LPFM service was severely limited by legislation supported by the commercial broadcast lobby. This legislation greatly restricted the number of LPFM stations that could be put on the air. Advocates of LPFM did not succeed in overturning that legislation until 2011, but even so 850 licensed LPFMs were broadcasting by 2010.[37]

Radio Cultures, Technical Cultures: Ham Radio, CB Radio, and Appropriate Technology

There are a variety of cultural antecedents to the form of radio activism explored in this book. People have been excited about the multidirectional "connectivity" of radio from its earliest days. Writing in 1926, just prior to the stabilization of commercial, networked broadcasting in the United States, German playwright Bertolt Brecht offered the following interpretation:

> [R]adio is one-sided when it should have two sides. It is a pure instrument of distribution: it merely hands things out.
>
> And now to be positive, that is to say, to turn to the positive side of radio, here is a proposal to give radio a new function: Radio should be converted from a distribution system to a communication system. Radio could be the most wonderful public communication system imaginable, a gigantic system of channels—could be, that is, if it were capable not only of transmitting but of receiving, of making the listener not only hear but also speak, not of isolating him but of connecting him.[38]

Thus Brecht encouraged people to imagine radio not as an artifact for listening to broadcasts but as an artifact for "tranceiving." By this he meant that radio could be used for two-way communication, though not in the sense of point-to-point communication such as telegraphy or telephony. He was suggesting that radio could collapse the distinction between receivers and producers of broadcasts, making everyone a listener and a speaker. These ideas strongly resonate with interpretations of radio promoted by LPFM advocates decades later. Consolidated broadcasting, although the dominant meaning of radio in the United States since the late 1920s, has not remained unchallenged; in harmony and discord with consolidated broadcasting, we hear alternate practices and interpretations of radio spanning decades.

Amateur radio operators, or hams, provided an interpretive framework and set of practices that constructed the ether as a site for multidirectional communication, allowing ordinary people to congregate and circulate messages. According to Susan Douglas, "hams have always insisted that listening in be an active, participatory pastime and that Americans always have a portion of the spectrum reserved for *them*—everyday people. They have demanded and cultivated a commercial-free zone in the spectrum in which individuals ... are allowed to transmit, to explore, and to connect with one another."[39] As documented by Douglas, the hams' presence in the ether, from the earliest days of radio telegraphy through the first acts of spectrum regulation in the 1910s to 1920s, shaped the practices that eventually became broadcasting and established the US definition of the spectrum as a public resource.[40]

Crucially, hams *tinkered* with radio technology, rather than simply using it to communicate. In the early twentieth century, congregating in the ether required one to build, maintain, and enhance the electronics apparatus hams used to transmit and receive. This legacy of hands-on work remained even as the practices spread and commercial kits subsequently became widely available. Electronics tinkering was a remarkably durable practice throughout much of the twentieth century, and radio kits and ham sets firmly anchored this hobby (especially among educated men and boys). Hams developed close relationships with radio technology and formed communities based on this technical affinity.[41] The radio activists in this book inherited tinkering from hams. This is not to say that there was a direct passage of tinkering from hams to activists, but rather that ham tinkering is a cultural antecedent to other hardware practices including those of these activists.

Such acts of cultural transmission are never perfectly straightforward. Although the radio activists owed an obvious cultural debt to the tinkering practices of hams, the politics they attached to tinkering largely contradicted the hams' form of engagement with electronics. As Kristen Haring argues, hams were often "resolutely apolitical";[42] they discouraged the use of the airwaves for discussion of politics and claimed that the logic of the technology lent itself to neutral, scientific rationality in radio communication.[43] In sharp contrast, the radio activists viewed radio technology as liberatory and promoted its use for political organizing (see chapter 6).

How did the radio activists wind up ascribing such a different cultural meaning to radio broadcasting and to tinkering? To answer that, we have to look to other cultural strands that influenced their enterprise. Citizens band (CB) was a later entry into radio culture, becoming available for mass use in 1958. Much to the dismay of the ham community, users politicized CB radio. CB's rise as a mode of communication happened to coincide with the ferment surrounding the civil rights movement. Proponents of civil rights (and later, black power) as well as segregationists adopted CB.[44] "Klansmen used their radios to better organize their racial terror activities by reporting to each other on the whereabouts of law enforcement or of their latest targets," writes Art Blake.[45] Likewise, African American groups used CB to coordinate defense of their communities and organize resistance. Taking up CB had could have overt political overtones, in contrast to the hams' ostensible neutrality.

CB and its user base possessed many features the ham community did not. Whereas ham communities were largely monolithic, composed of white middle-class men and boys, CB was far more diverse across lines of race, gender, class, and education.[46] CB was also forgiving technically in a way that ham radio was not: users did not have to be licensed, did not have to know Morse code, and were not expected to tinker with their equipment the way hams did. This drew derision from hams, who "sought a moral-technical high ground and disparaged CB operators as 'rule breakers' and '10-4 maniacs.'"[47]

Although CB and ham radio echoed Brecht's notion of radio as an excellent means for expression and two-way communication, these forms contrasted with each other in terms their relationship to technical knowledge. The radio activists combined elements of each of these antecedents in surprising ways: they embraced both technical know-how and a radically democratized user base for radio. They also hoped to democratize technical expertise itself (which seems potentially curious, because it is at odds with a legacy of tinkering as elite practice). Radio activists promoted technical skill without being technocratic or seeking the "moral-technical high ground" occupied by hams.

Looking beyond the boundaries of earlier radio cultures, American culture has a long tradition of self-reliance and small-scale technology, attainable by "everyman." Historical precedent can be found in the "republican gentleman" and "independent producer" ideals prevalent in the early nineteenth century.[48] But the Appropriate Technology movement of the 1960s to 1970s bears an even more striking family resemblance. Its adherents rallied around "small is beautiful" community-scale technologies and emphasized their potential for social and personal transformation. They cultivated a countercultural ethos, moving "back to the land," forming communes, and engaging in alternative lifestyle practices. And they emphasized kinship with nature and the environment in an explicit repudiation of technologies that were perceived to be massive, dehumanizing, and destructive (including nuclear weapons and power). A 1976 essay titled "Inner Technologies" claimed that "the present technological paradigm is clearly in need of replacement."[49]

Appropriate Technology preceded the period in this book by more than thirty years. The goals and values of the radio activists were not directly transmitted to them by the Appropriate Technology movement. Rather, this movement provided them with significant discursive and material resources for articulating the worth and meaning of small-scale broadcasting. That said, linkages between these groups proliferate, many of which are outlined in subsequent chapters (see especially 2, 3, and 4). These include *Sex and Broadcasting,* a visceral and irreverent book on community radio, the first edition of which appeared as a self-published handbook in California in 1975. Arranged as a collage of writings, some explicitly about radio and some not, a sampling of a few successive passages reads as follows:

> The moon of course is full and powdery—lying on its side, it balloons along the tops of the dark trees: by some mystery of perspective, the skies *wheel* in circles as we push through the night: the stars rotate as if we were racing time.... We have made time stop: no, we have made it run, through this ancient device of us in the long low wide-windowed rail car pushing aside the darkness, pushing the night into the bearings of the mysterious parts of our persons. (158, emphasis in original)

> There were themes which occurred and re-occurred during the three days of the [radio] conference. That of community radio stations keeping the doors open to the community; that of getting boring programs off the air; that of their relations to action groups which are striving to force commercial broadcasters into giving up time for blacks and chicanos and other minorities; that of the pitting of block programming (with a heavy, minute-to-minute program guide) against free-form or open programming.... We realized at that time that community, non-institutional radio was, all at once, young and active, and growing, and meaningful, and rich, and alive. (158)

> And there is the good feeling, the kameraderie [*sic*], among those of us who try to speak or think or act out of concert with the asphalt and concrete—the few of us in [San Jose] county who have found each other. (160)[50]

These passages encapsulate a few of the most salient aspects of Appropriate Technology: assertion of humanity in the face of an environment transformed by concrete; renewal at community and personal levels through fellowship, communion, and small-scale technology; and the reclaiming of a balance between humans and nature. Nearly forty years later, the radio activists I studied cited this book as a major influence: "Milam [its author] was one of the early pioneers in radical, community based non-commercial broadcasting.… At times the book reads like a yippie manifesto, but when Milam starts talking about the power of radio, one cannot help but be swayed."[51]

This quote about the "power" of radio demands scrutiny. Hams, radio activists, CB operators, media theorists, regulators, and, for that matter, yippies,[52] have all grappled with the utility and potential of communication technologies in general and of radio in particular, asking: in what ways can radio be understood to affect listeners, communities, or society? These issues are devilishly hard to pin down, and this framing is probably not an optimal one. As the example of CB illustrates, to generalize about the "impact" of this technology is futile (because it was adopted by Klan members, black power advocates, and truckers, for different purposes). Neither are technologies "neutral tools" whose uptake and effect are dictated solely by who is using them.[53] What the "power of radio" quote does productively point to, however, is that our understanding of media technologies is often quite flexible, even protean; here, the "power of radio," is understood to be profound, but is wholly unspecified. Is the "power" to be found in the ability to speak, the potential to disseminate and receive information, to achieve self- or community transformation, or the freedoms associated with technical participation and know-how? (We can assume that here "the power" is not to brainwash, stupefy, or propagandize, though in a general accounting of popular and academic understandings of media technologies, we ignore these at our own peril.) This book does not seek to answer what, exactly, the power of radio might be. Instead, it attempts to draw our attention to the range of ideation surrounding radio, including the fact that some of these notions may be contradictory. The vitally important question is how these notions get affixed to the artifact and with what consequences. The reason to survey the radio and technical cultures that precede twenty-first-century radio activism is that its lineage is complex. It contains multiple strands of thought and practice about why and how to engage with radio in the first place, including the prefigurative politics of transformation found in Appropriate Technology, the formation of technical community and identity embodied by hams, and the ability to address others for explicitly social or political purposes as embraced by CB radio operators.

Prometheus's Origins and Early Years

Prometheus rose out of the ashes of Radio Mutiny, an unlicensed broadcasting collective operating in West Philadelphia in the mid-1990s, which was shut down by federal marshals acting at the behest of the FCC in 1998. As one organizer told the story, the collective's commitment to radicalism and concern over having their unlicensed station receive external attention had prevented them from seeking too much exposure. They were more concerned with being a sustainable media outlet in their neighborhood than with being sensationalist, which might endanger them. But once they were raided by the FCC, the organizer claimed, "We called every news outlet in town and said, 'Hello. This is Radio Mutiny. Until yesterday, our station refused all contact with the mainstream capitalist media. But yesterday the Feds came and shut us down; we want to invite everyone to come out tomorrow to our protest at Ben Franklin's printing press. We are inviting the FCC to come arrest us at the site of this historic American monument to freedom of the press."[54] This narrative must be read as mythmaking. Often repeated, it served to portray the identity and origins of Prometheus in a meaningful way to themselves and others. Yet the radio activists did hold their protest and issued a challenge to the FCC: "for every station they harassed or shut down, we would teach people to build ten more."[55] They were staking a position and were poised for transformation (see figure 1.1).

With the FCC considering granting new community licenses, and Radio Mutiny shut down, some of the Mutineers decided that it would be worthwhile to take the government at its word and help community groups obtain radio stations legally. Part of the rationale for this was that they felt it was important to create stable new media outlets, rather than ones that would be in constant danger of being raided and shut down. Another was that they knew that they were bluffing to an extent; it would be taxing to travel around and actually build as many stations as they claimed they could. So they changed their name to Prometheus and took a seat at the table during the drafting of new LPFM rules. Soon after, the FCC announced what it called the "LPFM service." Prometheus then toured the country to talk to community groups and unlicensed broadcasters about this new service.

When I first encountered Prometheus, the organization had three full-time staff, which expanded to four and then to five during my fieldwork. In addition, they always had at least two or three interns and numerous volunteers coming in and out of the office, as well as occasional people working on contract projects for pay. The definition of who was or was not affiliated with Prometheus was fairly fluid, enabling some core members to

Figure 1.1
The radio activists protest the shutdown of their station at a statue of Benjamin Franklin's printing press, downtown Philadelphia (1998). An antenna placed by the activists rises above the press, which is also festooned with a banner reading, "I support pirate radio." Franklin himself has a speech bubble coming out of his mouth stating, "Necessity knows no law." Volunteer photo.

articulate the group's mission while others dropped in and out; I noticed this in many instances, including their willingness to name me as one of them when I traveled with them. (This affiliation did not come as naturally for me.) The paid staff was nonhierarchical throughout the period of my observation and formally adopted consensus governance after I was no longer observing, in 2007. When I began my fieldwork, the organization had recently added a board of directors to provide oversight and enable the transition of the group to 501(c)(3) (charitable tax-exempt organization) status. Board meetings were held approximately once a year, though some board members were in frequent contact with staff. I observed Prometheus's second-ever board meeting, held in early 2005.

Prometheus's 2003 lawsuit against the FCC (described below) propelled the organization to a more prominent stature in their policy field. A Prometheus organizer was quoted in the *New York Times* in 2003, which some in the organization saw as a milestone.[56] The organization also became more attractive to funders as a result of its higher profile. During the period of my fieldwork, its annual budget exceeded $100,000 for the first time. Organizers spent several months working to secure 501(c)(3) status, which they attained in 2005. Previously, the organization had been funded by a combination of private donations and foundation grants (including the Ford Foundation) facilitated by a fiscal sponsor. Workers occasionally laid themselves off and collected unemployment when funds ran low. The paid staff earned $10 an hour. In 2005 they began to toy with the idea of how to provide health insurance for the paid staff (an e-mail from a departing intern stated, "keep kicking ass and taking names. And I hope you get health insurance soon"[57]). These details underscore the scale, newness, and fragility of the organization during this period, as well as its trend toward growth and sustainability.

Because my fieldwork began in 2003, I have no direct knowledge of Prometheus's activities prior to that time. I have been able to construct a sense of the external issues they faced in the years 2000 to 2004 through interviews and observation of priorities while I was with the group. The major issues that emerged were the following:

1. Media consolidation threatened. In the Telecommunications Act of 1996, the US Congress relaxed rules that prohibited certain forms of media ownership and consolidation, which for radio meant that corporations were now permitted to own an unlimited number of stations nationally (as stated previously, Clear Channel Corporation had acquired more than twelve hundred radio stations across the country by 2001).[58] A public debate on media ownership, which was muted prior to the 1996 act, became much more prominent by 2003. Then-FCC chairman Michael Powell proposed new rules to permit even greater cross-ownership of media outlets. These rules would, for example, have permitted a newspaper formerly in competition with a broadcaster in the same market to be owned by the same entity. This proposal drew criticism from many quarters, well beyond organizations already devoted to media reform; such diverse groups as the National Organization for Women,

the National Rifle Association, and the AFL-CIO issued statements opposing the new proposal, and mobilized their constituents on the issue. The FCC received over two million public comments[59] opposing increased consolidation. Advocates claimed that the previous record was about 3,500 comments on the issue of low-power radio.[60] This was an issue of unprecedented public interest in telecommunications policy.[61]

In spite of the comments it received, the FCC went ahead with its proposed loosening of cross-ownership rules. The rules were supposed to go into effect in September 2003. Prometheus (with the aid of Media Access Project, a public interest law firm in Washington, DC focused on telecommunications issues) reacted by suing the FCC over the rules, appearing as the primary plaintiff in the case (*Prometheus Radio Project v. FCC*). This case was heard by the US Third Circuit Federal Court of Appeals in Philadelphia in early 2004, and to the chagrin of the FCC and the very powerful broadcast lobby, Prometheus won. As a result, the FCC was told to go back to the drawing board in June of that year. Media consolidation issues remained largely unresolved; advocacy groups kept watch for either a new bundle of pro-consolidation rules or a series of smaller measures.

2. Prometheus sought to increase the number of available slots on the FM dial for LPFM stations. When the FCC initially recommended the creation of the LPFM service, the National Association of Broadcasters (NAB) (the trade association for commercial broadcasters) and NPR opposed it, claiming that existing stations would experience "an ocean of interference" from the addition of the low-power stations to the dial. These claims were in direct contradiction of the FCC's assessment.[62] Nevertheless, in 2000, they succeeded in undercutting the FCC's initial recommendations for spacing between stations and were able to reduce the number of potentially available slots for LPFM by 80 percent. Advocates spent the next ten years trying to reinstate the FCC's original spacing recommendations.[63] LPFM turned out to be a more knotty political and technical issue than its proponents had anticipated.

In the political arena, the NAB made a concerted effort to challenge LPFM. It went to significant lengths to inform—some would say mislead—Congress about signal interference. It distributed compact discs in which the listener encountered unpleasant sounds, such as static, screeching, and incoherent overlapping audio streams, with the claim that radio audiences would be direly affected by the introduction of LPFM. NAB also championed The Radio Broadcasting Preservation Act of 2000 (HR 3439 and S 3020), which proposed cautionary spacing requirements for LPFM stations until interference that LPFMs might create could be studied. Though the bill succeeded in the House, it faltered in the Senate. Yet ultimately the new LPFM service was eviscerated when the Broadcasting Preservation Act's language was slipped into legislation as an appropriations rider (Commerce-State-Justice Appropriations). It became law when signed by President Clinton on December 21, 2000.[64] There is no doubt that many members of Congress who were receptive to the Broadcasting Preservation Act were genuinely concerned about interference when the issue was presented to them. Others may have also been motivated to support an interest group with deep pockets.

The audio engineering division of the FCC disputed NAB's allegations and went so far as to circulate in Congress a document challenging NAB's evidence:

> We are concerned that Members of Congress have received misleading engineering information about alleged interference from low power FM radio stations.
>
> One particularly misleading disinformation effort involves a compact disc being distributed by NAB that purports to demonstrate the type of interference to existing radio stations that NAB claims will occur from new low power FM radio stations. This CD demonstration is misleading and is simply wrong.[65]

A significant reason that the Federal Radio Commission (precursor to the FCC) was established in the 1920s was to give technical guidance to lawmakers whose expertise lay elsewhere. In this case, the broadcast lobby managed to enroll Congress to overrule the technical experts at the FCC who were themselves supposed to guide Congress. This underscores the unique political position of the technical experts at the FCC. Even though the bulk of accredited technical expertise favored the activists' position, they still faced an uphill battle as they pushed against a powerful lobby and a largely dysfunctional Congress to reinstate LPFM's original contours.

The seemingly arcane technical change in the regulation moved the spacing between LPFM stations and other stations on the dial from the "second-adjacent channel" to the "third-adjacent channel" (leaving three adjacent channels clear). This meant that within a certain geographical range and surface distance, a station could not go on the air at 91.9 FM if another station were spaced within 0.6 mHz; thus 91.7, 91.5, and 91.3, as well as 92.1, 92.3, and 92.5 were all unavailable. The FCC had intended to allow LPFMs to occupy spaces on the dial within 0.4 mHz of full-power stations because LPFMs' signals were understood to be weak enough that interference was unlikely; under those rules, 91.3 and 92.5 would potentially be available. (The standard spacing between two full-power stations is 0.6 mHz, and LPFM signals are far less potent.) But the December 2000 rider stipulated that LPFMs be kept 0.6 mHz apart from all other stations until potential interference could be studied.[66] As noted previously, this meant that about 80 percent of the stations that would have been available in the FCC's initial recommendation were countermanded.

The government retained the MITRE Corporation to study the potential for station interference. In July 2003, MITRE released its study, which indicated that the third-adjacent rule for LPFMs could be relaxed[67] and that opposition to LPFM on the basis of interference was unfounded.[68] However, this did not lead to the instantaneous reinstatement of the FCC's initial recommendations for spacing LPFMs on the dial. In fact, this goal remained elusive for advocates until 2011 because of continued political opposition by the broadcast lobby and a Republican-dominated Congress. This issue was still a key concern for the activists toward the end of my fieldwork in 2007.

The potential for interference was plainly not the only consideration for groups opposing LPFM. The permissibility of "translators" makes this especially apparent. Translators—transmitters that repeat the signal of a full-power station—are technically nearly identical to LPFM transmitters (actually they may broadcast at a slightly higher wattage, up to 250 watts). They are allowed to ignore the 0.6 mHz spacing if they operate below 100 watts. They are also permitted to use a less conservative terrain mapping methodology.[69] NAB did not contest the spacing requirements for translators, yet it objected to the recommendations for LPFM as being too lax. This raises strong suspicion that NAB was not genuinely motivated by fear of signal interference. Many LPFM advocates saw this opposition as motivated solely by a political agenda, based on fear of competition for audience and revenue. Paul Riismandel argues that "low-power FM stations have been permitted all along, but only for one class of owners: those that already own full-power stations ... since LPFM stations are also eligible to be placed in these spaces, they represent a small but crucial blockade to the established broadcast industry's growth into the last remaining open crevices in the FM broadcast dial."[70] The Local Community Radio Act of 2005, introduced in the Senate by John McCain (R-Arizona), Maria Cantwell (D-WA), and Patrick Leahy (D-VT), stated:

> Currently, FM translator stations can operate on the second- and third-adjacent channels to full power radio stations, up to an effective radiated power of 250 watts, pursuant to part 74 of title 47, Code of Federal Regulations, *using the very same transmitters that LPFM stations will use.* The FCC based its LPFM rules on the actual performance of these translators that already operate without undue interference to FM stations. *The actual interference record of these translators is far more useful than any results that further testing could yield.*[71]

The NAB also complained that it was worried about another form of "interference"—the economic impact on incumbent broadcasters, which was perhaps at the heart of the matter. The NAB was concerned that "audience diversion caused by additional LPFM stations would hurt existing radio service whether or not the new LPFM stations created [signal] interference."[72]

Had the MITRE study's conclusions been different, scientific controversy over technical "facts" might have ensued (advocates were braced for this). But with the bulk of the technical assessments in accord with one another, Prometheus and other advocates shifted emphasis again to Congress and the political domain.[73] Despite their efforts, LPFM failed to return to its initial status for over a decade due to the immense sway held by the broadcast lobby. The FCC was not a countervailing force on this issue. It was able to advise on technical matters and make recommendations based on its own (often shifting) interpretations, but it lacked the force of Congress.[74]

After 2000, and doubly so since the 2003 release of the MITRE findings, a major goal of LPFM advocates was to undo this legislation and restore the FCC's initial recommendations for LPFM. When I first met with Prometheus members in summer 2003, they were anxiously awaiting the release of the MITRE study and were wondering about the possibility of raising money to commission further engineering research if the study found their opponents' claims to have merit. They were very worried about the broadcast lobby's ability to "throw mountains of paperwork and studies" at legislators, believing that they would be at a disadvantage in terms of the sheer volume their opposition could effortlessly produce.[75] However, though MITRE's findings contradicted the broadcast lobby's claims, the NAB did not come around, instead decrying the study as "highly flawed."[76] This left advocates for LPFM feeling as though they had their work cut out for them in terms of organizing the grassroots while simultaneously running an inside game to line up allies in Washington, DC.

Each session of Congress was a potential opportunity to pass new legislation, and LPFM advocates and Prometheus consistently devoted time and resources to strategizing about how to attain this goal. The pre-2006 Republican-dominated Congress was not overwhelmingly receptive to this goal, nor was the FCC overly committed to pushing LPFM in Congress. Thus during the period of this book, the legislative arena in which LPFM advocates toiled was difficult for them to navigate. LPFM was becoming institutionalized, but it was at the same time far more precarious and less robust than advocates had hoped.

3. Prometheus's third major area of work was enabling new LPFM radio stations to be built. In 2000, Prometheus organizers "went on tour," driving around the country giving workshops about LPFM and showcasing the opportunity to apply for licenses. The FCC did not accept applications on an ongoing basis. It opened five windows (one for each of five geographical sections of the country) for applications in 2000–2001. If applicants missed these windows, they would be forced to wait until another window opened; none was scheduled for another twelve years (and in 2000, it was not clear another would even open again). The FCC received about 3,200 applications for new LPFM stations, around 1,000 licenses were granted, and around 650 new stations had gone on the air by the time my fieldwork was winding down in 2006.[77]

Prometheus's involvement in these stations varied considerably. On tour in 2000, activists taught people how to fill out application forms. They later served as a resource for groups filing applications. Prometheus members assisted a wide range of applicants, walking people through the application process and answering technical and legal questions to the best of their abilities. No one on the staff was trained as a lawyer, and only one person was an engineer (who joined the staff midway through my fieldwork). Despite this, Prometheus's exclusive focus on LPFM allowed them to gain quite a bit of technical and legal knowledge, though there were always highly specialized questions that caused them to call on or refer applicants to people with more expertise. The only type of broadcaster-applicant that Prometheus would not assist was the Christian broadcaster. They referred these applicants to another organization whose mission was to assist Christian community broadcasting groups. An instruction sheet hanging over the phones in Prometheus's office read, "We generally do not assist churches except in very particular cases—they can be referred to [christianbroadcasters.com]. No need to feel guilty about this—they will get better assistance there from someone who's [*sic*] mission it is to assist them. We are actually trespassing on [the other group's] 'territory' if we do services for Christian stations." (Chapter 6 offers more details on the uneasy alliances Prometheus and other left-wing groups forged with evangelical Christian and right-wing advocates for LPFM.)

The highest level of involvement for Prometheus in a new station was to collaborate with the licensees and plan a barnraising event (see chapters 2 and 3). Barnraisings were held over long weekends. Volunteers and Prometheus members descended on the site of a new station and worked to get the station up and running. Barnraisings often had between two and three hundred participants, including local and national volunteers, and Prometheus planned the event comprehensively in coordination with the local aspiring station. They provided meals, "crash space," home-stay lodging, and workshops to participants. These workshops covered programming, audio editing, interviewing, technical aspects of broadcasting, and station governance. Barnraisings also involved building and wiring together the computers, studio, transmitter, and an antenna. The work at barnraisings was quite varied, ranging from child care to cooking, cleaning, volunteer coordination, staffing registration, soldering, electrical wiring, software installation, carpentry, and setting up mixing and studio equipment.

The barnraisings were highly symbolic events where Prometheus performed and reinforced the twin missions of community radio and community organizing they envisioned. Prometheus would assist just about anyone who came to them for help. But the groups selected to have barnraising events all shared values with Prometheus; they were left-leaning and heavily involved in their local communities. A list of these organizations includes *Radio Consciencia,* the Coalition of Immokalee Workers' (CIW) station in Immokalee, Florida, owned and operated by a Central American immigrant farm laborer community heavily involved in campaigns for workers' rights (the CIW led the successful boycott against YUM! Corporation, Taco Bell's corporate parent, for unfair labor practices toward tomato pickers); WRYR ("We 'R' Your Radio") in South Arundel, Maryland, operated by an environmental justice organization in the Chesapeake Bay; and Radio Free Nashville ("Radio Free Nashville—Low Power for the People";[78] alternately "Radio Free Nashville—Unbuckling the Bible Belt"), a station operated by a group of politically progressive people in and around Nashville, Tennessee, who felt especially alienated from many residents of their city and desired a media outlet of their own. Prometheus usually held two or three barnraisings per year, completing a total of ten by 2006. The barnraising was deliberately set up to introduce and display Prometheus's mission, and was therefore an excellent site for interrogating the activities and artifacts of radio activism.

In the following chapters, I harmonize the more general analytic and historical contours of LPFM activism laid out in the book's introductory chapters with an ethnographic accounting. Prometheus was a small organization, and it may seem surprising that they played such an important role in media democratization. But the identity work and local activities of this small organization are important precisely because of how Prometheus was embedded in a broader political and historical set of developments.

2 Selfhoods: Geeks, Activists, and Countercultures

One Sunday evening in April 2005, in the fading light of a glorious sunny day, one hundred people sat on folding chairs facing a makeshift stage in a small town just outside of Nashville, Tennessee. A new LPFM station was signing on for the first time. With equal parts joy and catharsis, its founders recalled the obstacles they had faced: licensing setbacks, relocating their proposed broadcast site, and more immediately, a weekend in the mud with volunteers struggling to erect an antenna tower (the sunny day came on the heels of two straight days of pouring rain). A brother-and-sister team in their forties, Emily and Steve, had imagined the station and begun planning in 1997, so the sign-on was the culmination of years of effort. Activists and volunteers from Nashville and elsewhere gave speeches, which were followed by the performance of a radio play that barnraising volunteers had written over the weekend. After that, volunteer hosts went on air to introduce themselves and the programming they would offer on the new station. This included everything from Appalachian music to punk music to local politics.

The work and activity level over the weekend had been very intense, but even so the emotional experience of the inaugural broadcast was surprisingly overwhelming. Another volunteer presented Emily with flowers, and congratulated her on her "amazing birthing process," saying that although most women take nine months, Emily had taken eight years to bring her "baby" into existence. Having barely addressed the crowd, Emily began crying too hard to speak; after only getting out a few words, she kept choking up, and later she walked around the site smiling, crying, and embracing volunteers. Many people in the audience were also choked up or actively crying. A New York–based volunteer who had worked over the weekend to hook up computers and audio in the studio welled up in the chair next to me. (He said he was easily affected by public displays of emotion and always cried at weddings.)

The barnraising scene reveals a distillation of the radio activists' mission. People found each other—and parts of themselves—in the crush of activity that went into putting a radio station on the air. The event encompassed technical learning and the experience of fellowship and collaboration. At the same time, the barnraising potently sparked the imagination of a different world. This world was brought into being through the relationships between the new station, the people it served, and a movement that linked the Nashvillians to struggles for political inclusion and social justice in the wider world. This chapter explores the aspects of selfhood embodied in people's relationships with radio.

Radio activists not only interacted with technology but also formed close and complex relationships with it. This surpassed what many ordinary users of the same artifacts experience. Consciously and unconsciously, the radio activists projected political, affective, and aesthetic identifications onto radio technology. This meant shoring up associations between forms of selfhood and technology that they found desirable, as well as attempting to reconfigure identifications that had grown up around radio that they wished to contest. In all cases, technical practice served as a crucible of identity formation for the activists, community members, and would-be users to whom the activists promoted radio. Activists sought to demystify technology and thus to engage users politically.

How did the geeky, activist, and countercultural senses of self that are associated with radio form and become durable? This chapter shows the cultural linkages and performances of selfhood surrounding each of these identities, using barnraisings and a local gathering in Philadelphia (Geek Group, a pseudonym) as sites to excavate these processes of identity formation. More than merely constituting roles people stepped into, these forms of selfhood mattered because they became social tools that the activists drew on. They allowed activists to imbue activism and technical practice with the appropriate meaning. And yet there were consequences to the deployment of these selfhoods. Sometimes these identifications reinforced the politics of technology the activists sought to propagate, and at other times they presented uneasy contradictions that required reconciliation.

Mediators of technology are understood to be situated in between users[1] and producers.[2] They have often played a special role in assigning technology's meaning and prescribing its use.[3] Mediators are of special analytical interest because they have the potential to shape "upstream" and "downstream" interpretations of technology. This ultimately affects what is regulated into (or out of) existence, by determining how things are built, and how they are used. In this book I argue that the radio activists are "propagators" of technology. I use the term propagation specifically to refer to the intertwined practices of discursive and material engagement with an artifact. Propagators shape the form, meaning, and use of a given artifact through argumentation and mediating work to audiences such as users or regulators. They also build and diffuse artifacts and thus shape them materially. Thus, propagators are a special category of mediators—not only of use and interpretation, but also of technological engagement.[4]

The radio activists were not primarily interested in hardware per se. Though they labored to distribute technological artifacts and skills, they were also very concerned with the social relations surrounding radio. In the words of one Prometheus activist:

> What's important isn't the box that broadcasts the radio. What's important is the social structures that evolve.... [The reason we call the weekend workshops in which new stations are built "barnraisings" is that i]t used to be that if you wanted to build a barn, you had to be at least on speaking terms with your neighbors. You had to go out and say, "Hey, my barn burned down last week. Could y'all come out and help? We'll feed you." Now what do you do? You go in to your job, make a lot of money, hire a crane and they put it up for you. What these community radio stations do is bring people together. Those are the sorts of public community institutions that I think need to exist in order to counter the power that corporations have now.[5]

To this activist, *community* was more important than *radio* (chapter 7 engages this theme extensively). The barnraisings enacted by the radio activists were sites of tinkering, pedagogy and technical engagement. Although the activists appeared to orient primarily around technical artifacts, they possessed a more extensive agenda for social change.[6] Propagation thus involved tying artifact to politics and vice versa. The envisioned "output" of their technical engagement was not merely new radio stations with working hardware, but broadly democratic social relations, embodied in not only egalitarian technical practice, but also participatory media making and community self-determination.

Judith Butler has famously argued that identity is not something that is given, it is something that is constantly constructed and remade: "the 'doer' is variably constructed in and through the deed."[7] Identity is not endlessly fluid merely because it is performative or iterative.[8] Rather, identity is constituted through performance, materiality, practice, social relations, and signification. In my use of identity throughout this chapter and book, I take seriously the idea that identity is constituted through each of these means.[9]

Although some treatments of communication and information technologies use the concept of identity to debate or reify the boundary between modern and postmodern selfhood,[10] the use of identity as a space to debate the boundaries between modern and postmodern is an inadequate use of a category that can do other useful things. The benefit of using identity in the first place is to get at parts of human experience that are moving targets—slippery, constructed, and yet "real." Relationships with machines are well-documented sites of identity construction.[11] I conceptualize the radio activists not as people who simply worked with radio, but as people who actively constructed identities around their work with radio.[12]

The barnraising concept is a self-conscious reference to the Amish practice of people joining together to accomplish a project that an individual or small group alone would struggle to achieve, thereby emphasizing interdependence and cooperation. It was typical for Prometheus to organize two to three barnraising events per year. At a barnraising, about 150 to 300 people converged over a long weekend to hold workshops and build a new radio station. This included Prometheus staff, volunteers from Philadelphia, volunteers from the town and region in which the new station would be located, and some die-hard barnraising volunteers who traveled from all around the country to participate. A new station must go on the air within eighteen months from the station receiving its "construction permit" from the FCC (the final phase in the licensing process). By 2006 Prometheus had held around ten barnraisings with new LPFMs in the United States (three of which I attended during fieldwork). Prometheus members repeatedly stated that "a barnraising isn't the most efficient way to build a radio station, but it is the best way to build a movement."[13] The goal of the barnraising was broader than "merely" putting a new radio station on the air.

Geek Group was a gathering in Philadelphia that was similar to a barnraising in its emphasis on skill sharing, heightening identification with technology, and training novices. Though the group met in private homes, it was officially a project of the Philadelphia Independent Media Center (IMC). Attendance ranged from approximately four to fifteen people, and the meetups were less goal-oriented than a barnraising. Geek Group embraced tinkering for its own sake: they would often perform diagnostics on equipment that was beyond repair. Their work was usually related to FM radio technology, though other electronics projects were also considered within the purview of Geek Group (including building tone generators and synthesizers). Occasionally other general DIY technical or crafts projects were included, such as distilling alcohol (moonshine) or screen-printing t-shirts. When I began fieldwork in 2003, Geek Group was held almost every week. Toward the end of my fieldwork (two and a half years later)

Geek Group was more sporadic. By then, it occurred on an ad-hoc basis when the geeks wanted to undertake a specific project.

At barnraisings and Geek Group, the activists attempted to work out the meaning of radio technology for themselves and for participants; technology was a way to express activist values. In certain instances, the identities constructed around radio technology complemented one another, and in other cases, conflicts between identities emerged. Each of the identities constructed around radio technology were not necessarily in harmony with one another. These identities were not simply roles that people tried on; they served as resources for activists to shape the meaning of activism and of technical work for themselves and others.

Countercultural Identity

In the 1960s and 1970s, a social group Fred Turner terms the "New Communalists" turned toward communal living and selective adoption of technologies to enact their beliefs about social change.[14] The radio activists' heritage is, in part, the New Communalist or Appropriate Technology movements of the 1960s to 1970s. Both the New Communalists and the radio activists believed technical choices were political, and both conceived of technologies as tools for broader social transformation. By using the term *heritage,* I mean to imply that the radio activists were successors to some of the New Communalists' ideals and values, but they were not "descendants" whose values were directly transmitted to them by people who were active in the New Communalism or Appropriate Technology. Rather, these actors were (consciously or not) drawing on these past iterations of lifestyle patterns and attitudes toward technology; shadings of New Communalist or countercultural identity recurred in the radio activists. This was evident in everything from offhand comments to the deeper structural choices the actors made about their homes, their relationships, their occupations, and their presentations of self.

An instance of shared heritage with the New Communalists was the radio activists' communal living environments.[15] During the course of many months of fieldwork, I became accustomed to spending time at communal homes for Geek Group and Prometheus office-related meetings. The section of Philadelphia where the actors lived and worked was distinctively populated with communal homes, and it was common to identify where people lived by their house names, as opposed to their street addresses (examples include "House of the Divine," "Brick House," "Sandbox," "Situationist Front," etc.). Some were rental properties, and some were owned by the residents. One house inhabited by radio activists was a former squat that was later purchased from the city at a sheriff's auction. Some of the households in the area may have been turned into communal homes during the New Communalist era, but the ones I visited regularly were founded by much younger people in the 1990s and 2000s.

Similar to their predecessors in the 1970s, people in the communal homes I visited sometimes entered into nonmonogamous or nontraditional relationships. Occasionally, Jasper riffed on this by teasing Ellen, who was engaged. As her wedding approached, Jasper said he was going to tell the other people at the office not to make her feel uncomfortable about her "lifestyle choice," because he didn't want her to feel discriminated against. He said, "Even though some of us might not agree with her choice or choose it for ourselves, we have to make her feel accepted." This was a self-consciously humorous inversion of "sensitivity training" that might occur in a different workplace.[16] Barnraisings also had a reputation for fostering a free-love environment (albeit a low-key or subdued one). Although this was in no way the stated purpose of the gatherings, it was conventional wisdom among the activists and regular participants that people commonly "hooked up" at barnraisings. At a staff meeting before a barnraising, one organizer said, "We're totally radicalizing Free Press [a more mainstream media reform nonprofit, which was to provide organizing support at the barnraising], you get that, right? We're going to seduce them with our sexiness and our smell.[17] Everyone, [you have to] seduce one [Free Press] intern.... My whole [communal] house is going to the barnraising to get laid!"[18] This comment humorously underscores the Prometheans' countercultural position and values: they felt they could "radicalize" a mainstream nonprofit through seducing them (romantically and politically) at a barnraising.

There was a prominent anti-mainstream aesthetic to the actors' self-presentation. Some had tattoos, while others had dreadlocks. Many wore obviously old or secondhand clothing that had been altered with patches and stitching. A number of the men wore striking beards. Women often opted to not remove underarm and leg hair, and sometimes commented on this. These presentations of self were a way of performing an identity that was antibourgeois and defined by alterity from a mainstream.[19] However, here the aesthetic often deviated from that of the New Communalists. The activists owed a debt to the aesthetic and ethos of punk rock and DIY culture as well.[20] They commonly cited DIY as an ethic, emphasizing the ideal of people making things for themselves, free of reliance on paid experts or commercial systems. Though it was by no means the only music that the actors enjoyed or participated in, punk music was present in homes and the

Prometheus office. One activist even sang in a cabaret punk band. Another Prometheus organizer said his political awakening began as a teenager in the punk music scene: "I started going to shows when I was fourteen and picked up punk rock records and anarchist 'zines and picked up stuff about the scene and politics and DIY culture."[21] He spent time after college living in squats, working on activist causes, and "dumpstering" various necessities before taking a job with Prometheus. Although the punk aesthetic is also countercultural or anti-mainstream, it is different from the one embraced by the New Communalists. But there was some aesthetic overlap. Beards, in particular, gave Prometheus activists a superficial resemblance to back-to-the-landers (see figure 2.1).[22]

A very popular Geek Group event, which drew more people than the ordinary electronics tinkering events, was building a still and "moonshining" alcohol over a series of weeks. On another occasion, I encountered

Figure 2.1
Two Prometheus organizers with beards at a barnraising, foreground. The one on the right is holding an FM antenna. Pasquo, Tennessee (April 2005). Volunteer photo, courtesy PRP.

Jasper in his home breaking up blocks of frozen tofu with a crowbar. He said that this inspired him to fantasize about a cooking show he described as "Julia Child meets 'This Old House'—a cooking show with power tools."[23] Both of these instances illustrate the use of technology for small-scale, domestic projects that emphasize self-reliance. (The cooking show idea could also be read as a critique of a gendered division of domestic labor, fusing a masculine domain [power tools] with a feminine one [cooking].)

A back-to-the-land echo was occasionally explicit in the group. One morning, an office meeting was held at a home. Some people were cooking breakfast as others straggled in and began to consider the agenda. In the midst of this slightly disorderly gathering, someone who was just arriving pulled Jasper aside and said, "I just saw your cat outside eating out of a bucket. Do you want to stop him in case he shouldn't be eating whatever is in the bucket?" Jasper considered this and said, "Thank you, but I think the cat knows whether he should or shouldn't be eating.... He's very old and I don't see any sense in stopping him." Someone else then playfully piped up, "Don't stop him! He's probably starting a back-to-the-land movement!"[24] This comment was met with laughter and assent.

On another occasion, Jasper told a sociologist interviewing him about alternative media that "microradio was the solar power of the '90s."[25] This comment indicated that Jasper understood there to be a direct link between the propagators of these two technologies: some of the same people who had embraced solar and other kinds of sustainable energy sources were later drawn to microradio. Jasper himself held a BA in "Appropriate Technology" which he had self-fashioned at a liberal arts school in the late 1980s. The Philadelphia actors were mainly in their twenties or thirties, but barn-raisings attracted some older folks with engineering backgrounds who had long been engaged in a range of small-scale technology projects (including housing, energy, water, and of course radio). Some community radio stations have clear linkages to sustainable energy: KTAO, a community station in Taos, New Mexico, powers their transmitter using photovoltaic panels, and WJFF in New York's Catskills is a hydro-powered community station.[26]

Jasper believed that solar and microradio technologies had similar valences and could be considered "appropriate" for similar symbolic reasons. Solar energy was widely embraced by the Appropriate Technology movement for being cheap and accessible, simply maintained, and suitable for small-scale application.[27] Many of these criteria were considered applicable to LPFM radio as well. A Prometheus volunteer who was studying for a graduate degree in environmental science said in an interview, "I am interested in Appropriate Technology—capital A, capital T—for environmental stuff, and I don't think that I expected to find such a close analogue in the communication stuff, but the style and scale [of Prometheus's vision is similar].... Community is the only scale at which you can work effectively for the stuff that I'm interested in, sustainable communities, whether that happens through housing or energy or water...."[28]

Radio activists also found inspiration in a prototypically countercultural book called *Sex and Broadcasting: A Handbook on Starting a Radio Station for the Community,* first published in northern California in the 1970s. Similar to the *Whole Earth Catalog,* the book reads like "a cacophony of artifacts, voices, and visual design."[29] Early oversized editions were published by "Dildo Press," and the back cover reads, "To many commercial broadcasters, being found reading *Sex and Broadcasting* might be analogous to the discovery that Vice President Agnew uses *Quotations from Chairman Mao* for the first draft of his speeches."[30] The book's claims about radio resonate with the values expressed by New Communalists about technology supporting the development of individual—and community—consciousness:[31] "A radio station should not just be a hole in the universe for making money, or feeding egos, or running the world. A radio station should be a live place for live people to sing and dance and talk: talk their talk and walk their walk and know that they (and the rest of us) are not finally and irrevocably dead."[32] Its author, Lorenzo Milam, was a founder of KRAB radio station in Seattle in 1968. Jesse Walker describes the significance of KRAB:

> KRAB [was] something different. Weirder. Freer. It's something new: a wide-ranging forum for more species of music and opinion than most listeners knew existed. It's a kind of radio that values independence, irreverence, and creative, risk-taking, volunteer-based programming. It will come to be called *community radio.* But not yet. "Before that," Milam explains, "it wasn't community. The early KPFA ... and KRAB[33] stations were for the elite—those who wanted vigorous discussion, strong commentaries, shit-kicking interviews, and rich and controversial musical programming."[34]

Communalist cues about radio overlapped in provocative ways with countercultural interpretations of computing. Esmé, who over several years was an unpaid volunteer and a paid contract worker for Prometheus, described in an interview her background growing up in the Lower East Side of Manhattan in the late 1970s and early 1980s. Her father owned an arts gallery–computer store that was frequented by hackers, coders, experimental artists, and musicians. These people often engaged Esmé in conversations. She recalled that they would say to her, "Information wants to be free, and you've got to understand technology and you have to master it because otherwise people will use it against you."[35] She said that at some point, someone handed her a copy of *Computer Lib* by Ted Nelson. *Computer Lib* (subtitled *You Can and Must Understand Computers NOW*) was, similar to *Sex and Broadcasting,* a self-published book modeled on the *Whole Earth Catalog*[36] (see figures 2.2a and 2.2b). When it appeared in 1974,[37] it challenged the interpretation of computers as impersonal machines of bureaucracy. It instead heralded the PC, painting computers as tools for personal expression.[38] Esmé cited reading *Computer Lib* as a teenager as a formative influence on her attitudes about communication and technology. She also said, "How to use communication [technology] for warfare is really well-understood.... All kinds of technology, and I think especially communications technology, creates leverage and power. Often the thing [our society] understands best to do with [technology] is to oppress people."[39] This echoes the New Communalist sentiment about the transformative power of appropriating and using technology for personal and communal use in opposition to technologies as instruments of warfare, oppression, or dehumanization.

Barnraising events fairly vibrated with New Communalist ideals.[40] The stated goal of a barnraising was to accomplish the building of a radio

Figure 2.2a and Figure 2.2b

The covers of *Sex and Broadcasting* and *Computer Lib.* Although the former addresses radio and the latter computers, both are countercultural treatises on communication technology from the mid-1970s.

station and the fostering of "community" in order to build a movement to transform society.[41] Barnraisings emphasized interdependency and cooperation. They were usually held in small-town or rural settings (largely due to the fact that LPFM licenses were not available in urban areas during 2000–2011). People often camped or slept outside or were provided with home-stays, which lent a feeling of constant togetherness to the weekend. At a barnraising in Oregon, a vacant and possibly condemned house near the work site was appropriated for "crash space," not unlike a squat. This space certainly fostered "togetherness" and intimacy amongst the people staying there (mostly younger folks). I found the lack of privacy and quiet stifling and somewhat regretted not signing up for a homestay with a volunteer the way I had at other barnraisings.

Mealtimes were structured for everyone to take a break from workshops and labor in order to come together to eat, talk, and relax at the work site. At one barnraising, the food was mostly donated by local restaurants that were enthusiastic about the new station and its politics. At least one meal was also provided by Nashville Food Not Bombs.[42] All were provided to attendees without additional charge as part of the sliding scale registration fee. I had initially thought I would use mealtimes for informal interviewing, but often found myself needing a break from all the frantic activity. At one memorable meal during the Nashville barnraising, a handful of volunteers sat in near silence in a room with the door shut (when people opened it, they were welcomed but told only half-jokingly, "Introverts only"). Two of the local men quietly discussed the similarity of the social space they had created in the room to a Quaker meeting. One said that he saw himself "as guided by [his] own inner light."[43] The barnraising enabled otherwise dispersed people to step out of their usual routines and spend time communing and reflecting with other like-minded folk. Though secular, an element of something akin to religiosity was palpable in the fellowship at this gathering and others.

Geek Identity

Geek first makes an early appearance in the English language referring to circus performers. (These performances featured such freakish acts as biting the heads off chickens.) Later, the meaning of "geek" drifted to signify people who were freaks about scholasticism or technology. The *Oxford English Dictionary (OED)* defines *geek* as: "*depreciative*. An overly diligent, unsociable student; any unsociable person obsessively devoted to a particular pursuit." It lists a more recent definition as "a person who is extremely devoted to and knowledgeable about computers or related technology," noting that "in this sense, esp. when as a self-designation, not necessarily *depreciative*."[44] Because Geek Group participants referred to themselves as geeks in their gathering's name and elsewhere, I retain that term in my discussion. Though the word originates as a term of insult, these people (and many others) use it as a fond, self-aware term to describe themselves and others. Geeks have laid claim to a word associated with a history of disparagement. They derive strength from its use to highlight their own uniqueness from others and their commonality with each other.[45]

Geeks may or may not differ from nerds. In his study of geeks, nerds, and race, Ron Eglash conflates the two, noting that the amount of writing devoted to distinguishing between them is, to him, "surprising." He quotes novelist Douglas Coupland's claim that "a geek is a nerd who knows that he is one."[46] In the case of these radio activists, I find *geek* to be of more value, due to its specificity, its relationship to technology,[47] and the fact that activists used the term in the self-aware sense noted by Coupland. Geek-nerd identity has historically been a means by which access to technical cultures has been restricted, especially along lines of race and gender.[48] But in the radio activists' conception, geek identity was intended as a means of *opening up* access to technical cultures. They tied technical engagement to political consciousness and intended technical expertise to be attainable by all.

Similar to radio hams (who display "technical identity" in the words of Kristen Haring), these activists held a closer relationship to technology than average users.[49] Here, geek identity was not only linked to technical skills per se. It was also displayed in arcane knowledge about a range of topics. Many of the geeks were quite self-aware in embracing a sense of self that related to their technical skills and to other activities that are considered "geeky": Ellen, a Prometheus staff member and occasional Geek Group attendee, volunteered to me that she and her housemates enjoyed role-playing games. One evening at Geek Group, the activists teased me, calling me a "meta-geek" for studying geeks. Coming from them, I could not help but take this as a compliment. The "meta-geek" comment may be read as an indication that I was welcome and included as far as the members of Geek Group were concerned. Even if I couldn't diagnose what was wrong with a transmitter board, I still had something in common with the group, as I was a different species of geek. Echoing Coupland, a good deal of what constitutes geekhood is an awareness and embracing of one's identity as a geek.

Geek Group usually alternated between two residences. Activities at the gathering typically consisted of troubleshooting electronics equipment, thinking about how to solve technical problems, and building equipment such as antennas and transmitters for use by Radio Goldman or other groups. (Similar to Geek Group [and also pseudonymous], Radio Goldman was a project of the Philly IMC. During my fieldwork, they were limited to webcasting, though prior to this they had shared airtime with a local 10-watt Class D station.) Much of the activity involved looking up, building, and testing antenna designs, beaming signals across the neighborhood from rooftop to rooftop, or troubleshooting instruments that didn't work properly (the geeks had a lot of scavenged and secondhand diagnostic equipment). Sometimes, they were just playing around, as was the case one night when Rolf was downloading large files from the Internet using a wi-fi connection while Simon picked up the signal on a spectrum analyzer and amplified it through a speaker, listening to the data transfer and pauses.

A lot of Geek Group's hours were whiled away in basements and on roofs. Basements were where workbenches and soldering stations tended to be located. Roofs had the advantage of elevation, which was important for signal transmission, whether wi-fi or FM. At the same time, these spaces within the home were removed from core domestic spaces such as kitchens or living rooms. This division of space is very similar to how ham radio operators carved out private territory (often the "ham shack") distinct from home spaces shared with or controlled by women.[50] Roofs in particular held a certain allure. They offered a perspective on home and street that was slightly unusual and to use them for hobbyist projects almost felt like claiming neglected space (perhaps especially so in the rambling dereliction of much of residential Philadelphia). One warm summer evening, Jasper and I waited on his roof for others to get into position on another roof; we hoped to be able to receive their wi-fi signal. While we waited, he spontaneously decided to use the time to nail down some shingles on the side of his house. To do so, he leaned off the edge of the roof, asking me to hold his legs and brace myself on the roof so he did not fall. This required us to sustain physical closeness. It caused me some anxiety because for obvious reasons, I did not want to slip or lose my grip on him. This episode illustrates a number of mundane yet telling aspects of this kind of work: improvised tinkering, an emphasis on repair, and immediacy between bodies and tools. Like much of the desultory work of Geek Group, the experience on the hot and dirty roof—waiting for Jasper to wrap up his repair or for our fellow geeks to make contact—felt alternately meditative and boring.

The basement of one Geek Group house was piled from floor to ceiling with carpentry and electronics, PCs, radio equipment and other electronics, lumber, and cables. The house upstairs was also often cluttered with pieces of computers and miscellaneous electronic components. It was home to a musician who composed and built instruments for electronic music pieces. Another resident was a Radio Goldman member who worked for the university's computer science department and often brought home cast-off equipment from the engineering school. Novels by popular science-fiction writers William Gibson, Philip K. Dick,[51] and Neal Stephenson were flung about the living room and kitchen, as were books on media theory and philosophy. Another Geek Group participant said that his bedroom in his own home was so cluttered with PC shells that he used them for furniture.

Geek signifiers abounded. One summer night Geek Group hosted screened two films: an obscure pre-*Star Trek* movie starring William Shatner that was filmed in Esperanto, and *Forbidden Planet*, a 1950s sci-fi film that is generally credited with having the first all-electronic movie score. Another evening, while looking up some information about building directional wi-fi antennas using a Pringles can ("cantennas"), Rolf was absentmindedly whistling to himself; I recognized the tune as the Death Star motif from the *Star Wars* films. Another geek said that he had recently taken up knitting and that one of his most satisfying projects was knitting a hat in the shape of a Klein bottle (a closed, continuous shape with one surface, like a Mobius strip). Another person wrote radio plays, one of which was about the early history of radio, a drama that centered on Nora Stanton Blatch, a feminist activist and Cornell University civil engineering graduate who was married to Lee DeForest and worked with DeForest in his laboratory.[52] A few of these activists collected and circulated "Radio Boys" books, a serial from the early twentieth century similar to the Hardy Boys that emphasized virtues of masculine technical competence and boyish adventurousness.[53]

There are some resonances between these radio geeks and computer hackers, especially in the strong, affective, even intimate relationships with technology that hackers and radio geeks both forge.[54] Yet many hackers value control, mastery, and virtuosity. In this regard, the radio geeks differed from, and even challenged, the hacker ethos (which historically has also been overwhelmingly masculine). Sherry Turkle writes, "[Hackers'] mastery [of] games and initiations test the ability to win over complexity and break out of confining situations.... [In hack situations such as lock picking or code breaking, t]hey are not after material goods, but after the thrill of triumph."[55] Though she mentions that for some hackers, willingness to defy the establishment may be tied to a diffuse notion of being an "electronic Robin Hood,"[56] the politics of the hacker ethos she describes are far less considered than those of these geeks.[57]

By contrast, the radio geeks were committed to raising political consciousness above all else; they did not value technical virtuosity in and of itself, placing them at odds with many strains of geek culture. Radio activists viewed the acquisition of technical skills as tied to demystification of technology. But technical skills alone were never enough to enact this consciousness. Rolf said he instead perceived what he called a libertarian or technocratic streak in many people with technical skills, which was opposed to the value of radical egalitarianism.[58] Although some people may distrust the government or corporate power, Jasper claimed that "people who become experts often come not to believe in democratic decision-making processes."[59] Another radio activist specifically mentioned hackers' technical skills and nascent antiestablishment sensibility. He suggested that activists could do outreach in order to guide the hackers to put their dissatisfaction into a more politically sophisticated or useful framework. For the radio geeks, technical virtuosity without political consciousness was limiting, and technical virtuosity alone was insufficient for the formation of geek identity.[60] In 2006, one of the activists sent me a text message to tell me that she was on a trip in Michigan where she had met "the worst geeks ever," and she wanted to talk about them later. In an interview, she said: "[Those geeks] specifically do not care about building tools for society.... They don't understand the intersection between technology and society, or how to use technology to be a better person ... [This] is disgusting behavior!"[61]

Both hackers and the radio geeks are motivated by the pleasure they take in their activities: Pekka Himanen writes that the hacker "finds programming intrinsically interesting, exciting, and joyous."[62] It is worth considering the fact that the "work" that occurs at barnraisings and Geek Group is *fun* for some people.[63] As mentioned previously, sometimes during downtime between "productive" activities, the geeks were simply playing, as in the example of Simon listening to the data transmission while needed files were downloading or making silly puns as they worked. (A geek word play: One evening, one of the men who still went by his old "pirate" handle said that sometimes people who did not realize that this was not his given name would ask about the ethnic background of the name, and he said that people most often guessed it was French. Simon exclaimed, "You should tell them it's Ancient Geek!") On some level, most people who attended Geek Group came to enjoy themselves. The incentive is plainly higher to spend one's free time doing radio tinkering if one considers this a gratifying activity. The geeks' activities, such as film screenings and moonshining, were pursuits that were not directly in service of the "needs" of the group to produce working radio hardware. They were "fun" events aimed at a wider array of people in order to forge a sense of commonality.

A fundamental premise of barnraisings (and Geek Group gatherings) was to learn from one another and share skills in an open, participatory manner. In theory expertise was to be leveled or at least distributed as equitably as possible. Activists hoped that when outside volunteers departed a barnraising, the new station would have enough comfort and expertise to run every aspect of the station. Activists also hoped that volunteers unaffiliated with the station would have experienced learning from and teaching others. One volunteer described the barnraising model in Prometheus's newsletter:

> The main thing I like about the barnraising as a model is that it accomplishes many things simultaneously. Prometheus could just hire a couple of builders and contractors, and have them come in and put the studio together, do the computer and electronics work and be assisted by a couple of local volunteers and have done with it [*sic*]. That this is not the path that Prometheus chooses is instructive. The intent is for the barnraising to be educational ... and at the same time actually produce a concrete physical functional thing at the end of the process.[64]

Identity formation occurred at barnraisings. The goal of the Prometheus organizers was an explicitly political one: to engage people in a movement to take control of media in order to enact widespread social change in service of social justice ideals. The link to technical practice may not be immediately obvious. The idea was to give people enough comfort with technical expertise to help them to see a link between technical expertise and other forms of elite knowledge. Organizers believed that if people felt less intimidated by technology, they would come to question assumptions about the necessity of elites making decisions for others. The goal was to technically empower barnraising attendees to encourage a radical, democratizing effect on decision making. This specific iteration of geek identity rested on both the creation of an idea of hands-on accessibility (including some comfort level with technology) and also on the creation and maintenance of political consciousness.

In practice, the barnraisings involved many opportunities for technical hands-on learning and participation. Standard activities in which volunteers and novices were encouraged to participate included soldering, field audio recording and production, and antenna building and tuning. Workshops taught people about the basics of how radio works. A popular workshop was "Introduction to Radio, Transmitters, and Antennas."[65] There were ongoing "work tracks" that people could join or observe, which were separate from the more formal workshops. Unlike a workshop, which lasted an hour or two, the work tracks were simply work that was ongoing throughout the weekend onto which volunteers could attach themselves. Some work tracks were very easy for volunteers to plug into. Activists set up a transmitter-soldering work track at the Massachusetts barnraising in order to encourage volunteer participation. These transmitters would later be taken to Prometheus's next station-building event in Tanzania, but they were not needed for the Massachusetts LPFM's operation (see figure 2.3). Most work tracks, however, were oriented toward work that needed to occur for the station to go on the air. Activities included setting up a working studio-transmitter link, carpentry to build the studio, setting up the station's computers with the appropriate software, and tuning and positioning the antenna. Certain people were designated in advance of the barnraising to be in charge of making sure that work happened on each track. Activists usually set up target times by which certain tasks need to be completed, but the structure for volunteers to be incorporated into the work tracks was

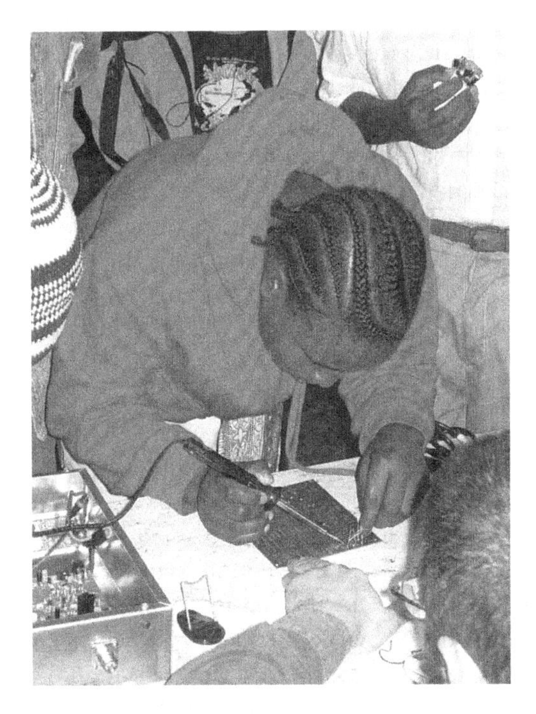

Figure 2.3
A Philadelphia-based radio activist teaching a young woman to solder while others look on at a station-building event in Imbaseni, Tanzania (August 2005). Author photo.

largely left open to the individuals in charge of the work. Prometheus did not train workshop or work track leaders or supervise their pedagogy in any formal way. Prometheus members did, however, repeatedly stress the ideal model of teaching and skill sharing. For example, during the antenna tuning in Tennessee, Jasper swooped in to chastise the engineers. He told one of them to "give the screwdriver and wrench to someone else!" He wanted the volunteers to use the tools. But unlike the workshops, the work tracks were arguably more about getting necessary work accomplished than providing hands-on experiences to novices.

An activity such as soldering a transmitter board, however, was a good opportunity for novices to participate. It was relatively simple, because the boards come with instructions and can be assembled in a "paint-by-numbers" fashion (as long as the instructions and schematic are closely followed). Soldering a transmitter board takes several hours of work and is a social activity. Each small board can accommodate a couple of people soldering and at least a few more observing or guiding the process. The parts are differentiated enough from each other to enable the instructor(s) to point to them and converse in general terms about their roles ("See the components that look like lentils? Anyone know what they are? Capacitors? What do they do in the circuit?"). There is a small switch that is used to set the frequency. There are parts that light up when the power is switched on (diodes). When the board is complete, if it has been assembled correctly, the instructor and volunteers can enjoy the fruits of their labor by testing the board and hearing that it works. This is done using the transmitter, a portable receiver, a power source, and an audio source. Over many months, I came to recognize a Manu Chao CD because it was the default audio source in a Discman that traveled in Jasper's toolbox; if Manu Chao was audible over a receiver, the transmitter being tested was working. The test was often turned into a theatrical event ("We'll test the board at 9! Tell the other people who worked on it to come by!"); people who only soldered for ten minutes would stop by later either for the initial test or afterward to turn the transmitter on and hear it work. It should be noted that the transmitter-soldering workshops at barnraisings were more symbolic and pedagogical than instrumentally necessary: the transmitter the new LPFM station would actually use would usually have been bought preassembled, so the ones built by volunteers were not in fact needed for the station to go on the air. Thus the transmitters running at LPFMs were more "black-boxed" than the activists' stated ideal. The lower-wattage transmitters built by teams of people in the workshops were put to other uses, including being given to people overseas.

Barnraisings were labor-intensive, with work occurring simultaneously at many sites. It was generally possible to circulate, stop in at stations where work was occurring, ask a few questions, and either move on, work for a few minutes, or work for hours. At the Tennessee barnraising, in addition to attending workshops, I made multiple errands to a hardware store for last-minute supplies. I also stayed late into the night one evening helping a Philadelphia-based volunteer custom-build shelves for the studio equipment. I observed the antenna tuning and helped to tune it; Queens, New York– and Illinois-based engineers led a group that did this on Saturday afternoon. I made turnbuckles for the antenna tower's guy-wires, and I staffed the registration desk for an afternoon. On another day, a New York–based volunteer and I ran electrical cables from a neighbor's house into the main tent, where there was to be dinner and a plenary session that evening. The neighbor had initially been opposed to volunteers setting up tents, camping, and tromping across his property, but he eventually relented. Unexpectedly, he also gave permission to plug into his electricity when he saw how much trouble we were having getting enough power from the property where the new LPFM was located. We had to be careful not to overload the circuits and we had trouble keeping the lights on. We also had difficulty with the massive amounts of mud generated by rainstorms during the event. Similar to the experience on the rooftop, this activity was also dirty and plodding. We did eventually get the lights set up.

This email from a Prometheus organizer named Renée encapsulates the goal of the barnraising:

> [The barn-raising strategy is] the idea that hands-on participation in building the station has a transformative effect on how members of a communitiy [*sic*] feel ownership of the media in question.... this moment for me is captured in [a film of the station-building event Prometheus conducted in Tanzania] when the young man is explaining—in swahili??—how the contraption (aha!) he has made works. then he steps back, looks at the camera, and says in triumphant english: *"and *i* made that!"*[66]

The "contraption" to which the organizer refers is a transmitter board. As stated previously, there are particularly compelling reasons to use transmitter boards to introduce novices to technical work with radio. Her statement underscores the connections of technical expertise, political empowerment, and political change central to the geek identity that Prometheus embraced and hoped to promulgate.

Brian, another activist, reflected in an interview about Prometheus's model for expertise sharing. He lamented the difficulty of implementing Prometheus's vision under the real-life circumstances of a barnraising. A major feature of barnraisings was the flurry of activity. There was typically an overwhelming sense of the need to accomplish certain tasks by certain times in order to stay on schedule for the ceremonial Sunday evening on-air event to close the barnraising. But Brian was also serious about trying to facilitate learning and hands-on engagement with technology. He said that he felt constant pressure about getting the station up and running versus fully bringing novices up to speed technically and fulfilling the Prometheus vision of empowerment through technological skill and identification: "I wish there were more time because there's so much explaining of the way things work [to be done] ... one of the ultimate failings of the barnraisings is the lack of time, you don't give people a good understanding of how something totally works even if they may have soldered some cables.... I kind of regret that stuff.... I have to make sure that the project gets done with an inclusive mission at the same time."[67]

Brian then broadened his scope to comment on expertise and expert-lay relationships more generally:

> What I hope people can take from the barnraising is that they can extrapolate the DIY attitude to everything else in their lives. I don't explicitly say that—maybe I should—but this isn't like a unique thing. You can do any tech project, or it doesn't even have to be a hands-on building thing.... Culturally we have a very expert-oriented society, where you have people on the news being interviewed about the political situation in Iran, or the stock market, and you have all these people who are "experts," and just because they're talking at you about these different things, doesn't necessarily mean they're right.... I've come up with a number of different ways to get people to just put their hands on the tech, just try it. Once they try it, it's not that complicated.[68]

But technical practice was still the central way the Prometheans tried to make the connections between technical demystification and challenges to expert power. They wanted to build participatory expertise: "The big part of the barnraisings, about not having the engineers do it, it is a demystification, and making people feel like, 'oh, experts just happen to know this, they've just done this a bunch,' giving people the feeling, 'oh, if I just did this enough, I could do this just as well as this guy, as well as this engineer.'"[69] Brian concluded by reiterating that this was always extremely difficult in practice.

Both organizers' statements illustrate that Prometheus sought to *create* in barnraising participants a political consciousness that challenged elite models of expertise through hands-on work with technology.[70] When Brian said that he hoped people would realize that the main difference between themselves and engineers was that "[engineers] have just done this a bunch [of times]," he was pointing to a process similar to the one Renée previously highlighted. When the novice young man said, "and I made that," the organizers saw a layperson having an awakening about technology, in which it is no longer the province of experts. Geek Group and barnraisings were sites at which geek identity was promoted, whether or not technical skill was significantly enhanced. Geek identity was a means of building and sustaining identification with media democracy activism.

Activist Identity

When asked about their political leanings, these actors universally identified themselves as "garden-variety lefties and" "leftist activists" who were "committed to social justice." Only a couple of them mentioned anarchism as an explicit ideal, but the group clearly identified with radical politics. Just one example of this was the name of the Philadelphia IMC's radio station, Radio Goldman (named for feminist anarchist Emma Goldman).

Many radio activists viewed their work with LPFM or community radio as integral to their broader activist agenda. They spoke of media activism as being a central tenet of social justice work, feeling that local, diverse media was important to community self-determination. An activist named Rose said that she first learned about microradio in 1997 at an activist conference on media and democracy in New York City. There she heard a presentation by a New York City pirate station, Steal This Radio (see figure 2.4):[71] "They did a presentation at night in this big auditorium with the lights turned down so we couldn't see their faces, it was like you were listening to the radio only they were right there.... It was all so exciting and I felt I was in the middle of a revolution.[72] Immediately afterward, she began volunteering with the Mutiny collective in Philadelphia, which was engaged in unlicensed broadcasting. Other radio activists felt that radio activism was an end in itself. One claimed that "the airwaves belong to the people, and the FCC isn't doing its job." He also claimed to be a "free speech absolutist" who he often felt like a "rogue" in the Goldman group because he was more committed to free speech than to normative leftist ideals.[73]

A useful concept for thinking about the formation of activist identity is the idea of "critical agency."[74] "Critical agency" implies a space within or outside normal social relations from which knowledge about those relations can be articulated. This allows for the ideation and voicing of critique. Activists tend to believe that they have critical agency. They may even *need* to believe they have critical agency in order to articulate their positions as ones of dissent or alterity. In the case of the radio activists, identity work occurred around this critical stance. Their work with radio was perceived

Figure 2.4
Steal This Radio, 1990s pirate radio in New York City. The station's name invokes Abbie Hoffman's countercultural classic.

as a direct application of critical agency. In the activists' conception, work with low-power radio was a particularly viable way to express criticism of mainstream, corporate media. Building alternative media was seen as a path to empowerment and would support a wider social justice agenda.

Leftist activist identities were in some ways complementary to radio geek identities. Activist and geek identities involve a celebration of being outside the mainstream; they are defined in opposition to this mainstream. Some of their activities give a sense of this. Scavenging cast-off electronics from the university delighted them. It was a performance of the rejection of the mainstream value of consumption. Reusing this old equipment required technical engagement in order to identify components and ascertain which were worth taking home. Jasper ruefully noted after one of these sessions that he had seen ultrasound equipment which he thought would be great to grab for "midwife or DIY abortion friends." Unfortunately, it had been missing the transducers (paddles), which he said were expensive to replace, so he opted to leave it in the trash.[75] On Bastille Day, the geeks held a celebration that involved fireworks and antiestablishment revelry. There was a small pageant of remembrance of the storming of the Bastille and the freeing of prisoners, celebrating anti-elite values and overthrow of current social orders. The event also involved a somewhat technical project that invoked the Gas Laws to launch hot air balloons made from dry-cleaning bags and Roman candles. Both scavenging electronics and celebrating the storming of the Bastille with DIY thermodynamics illustrate how critical agency and geekery were intertwined in radio activism.

Brian, a staff Prometheus activist with a background in computer engineering and electronics, discussed how his work at Prometheus with radio activism had for the first time enabled him to successfully focus and integrate his technical skills and his strong interest in social justice. He indicated technical practice could be tied to social justice goals. For him, this was the most meaningful application of technical work:

> [The activist work I was doing outside of my schoolwork, combined with the engineering studies in school,] created this sort of internal problem in my life. The things I was doing in school weren't in any way related to the things I was doing outside of school. It was like two separate halves of my life, and it just felt real weird....
>
> I had ideas of, oh, maybe I could do pirate radio stuff, but it turned out that the organizing side of pirate radio was much more involved than the tech side of stuff, once the tech side is set up, you're done.... I have not only a good working knowledge of computers and electronics, but how to apply them.... I never really got it together enough to ... do that stuff exclusively until I started working at Prometheus ... [where] I could just come on and bring my skills to something, [I] finally realized what I wanted to do was this, was to help people figure out how to make technology work for them....
>
> And radio is a great [choice], because it's simple, and it has a really direct effect, and a short-term effect, it shows effects pretty quick, the media is just so immediate. It's been really great to finally feel like I'm finally doing what I'd been working towards for so long.[76]

For Brian, the challenge of tying technical know-how to political or social goals was great. He felt that his work as an organizer at Prometheus enabled him to harmoniously accomplish these goals.

The legacy of pirate radio (with which only a few of the 2000s activist group had ever actually been involved) still carried on in spirit. This represented an iteration of activist and geek values. In one session of Geek Group, a portable transmitter the group was working on was screwed into a tin lunchbox painted with the Jolly Roger. This was left over from the days of the group's unlicensed broadcasting (see figure 2.5).

Goldman went off the FM airwaves in 2003 after a conflict with the community station whose Class D license they were sharing. When this happened, they received an anonymous e-mail from pirates who said that they would still be broadcasting Goldman's web-stream over FM. Goldman's official stance was that although they strongly valued FM, they were committed to broadcasting legally. Nonetheless, they were pleased to hear this from the pirates. A Prometheus activist and former pirate said his personal legal standard was, "will a judge laugh at you?"[77] He did not condemn

Figure 2.5
A portable lunchbox transmitter (2005). Note the Jolly Roger and anarchism iconography on the t-shirt of the person holding it. This was not a Prometheus transmitter, but one built for emergency communications in New Orleans after Hurricane Katrina. Courtesy Bradley Stuart Allen.

pirate activity. He was merely being circumspect about how closely Goldman or Prometheus should entangle themselves with pirate activity if they wished to retain their statuses as legitimate organizations. A close affiliation with pirate broadcasting would make it difficult to hold open the possibility of being involved with licensed broadcasting in the future, should the opportunity arise. This was a particularly real concern because the FCC had attempted to deter microbroadcasters by pledging that it would deny them licenses in the future if they were known to have participated in unlicensed broadcasting.[78]

The activist who told me about the email to Goldman from pirates said that pirates would continue to broadcast regardless of the potential of legal access. Some pirates felt that the airwaves were inherently public and they need not apply to the government for the right to broadcast;[79] they also perhaps valued doing something extralegal related to activist beliefs and self-expression. Pirate FM broadcasting was a site in which technical skills were used in the service of activist or antiestablishment beliefs. One former pirate said, "I did the pirate radio thing for about a year; it was a chance for me to hone my [technical] skills and play around and, you know, fight back at The Man."[80] Interestingly, a couple of people also expressed the idea that pirates possess the "best" technical skills. The logic was that pirates' technical skills and solutions were borne of necessity and therefore resulted in greater ingenuity. True or not, this rapturous attitude toward pirates suggests their elevated position in the minds of the radio activists.

The barnraising events were explicitly designed to be activist political events. Prometheus only organized barnraisings for groups whose missions they deemed to be worthy of their energies. The groups Prometheus selected tended to have a strong progressive vision and to serve in their community beyond being nascent radio stations.[81] They were often marginalized by geography, economics, or ethnicity. And they had a real need for Prometheus's help: the groups were in some doubt of having the expertise and human power to build the station without some assistance. The radio activists envisioned the barnraisings themselves as opportunities to build a movement around community media, with the goal of effecting widespread social change. The activists routinely forged connections between the media democracy movement and other struggles for social justice. The plenary sessions at barnraising very much resembled political rallies. Barnraising workshops covered a range of topics, including political ones, such as "Radio and Labor Organizing" and "The Movement for Media Reform" (see chapter 6).[82]

At a 2006 barnraising for a Latino farmworkers' union in Oregon, the connection between the power of organizing in social justice struggles and the barnraising was explicit. At a plenary session prior to the station going on air, a union organizer led the crowd in a "campesino cheer:" he directed the crowd to clap together their pinkies, then their pinkies and ring fingers, and so on, until gradually the crowd was making thunderous applause through the united force of everyone's ability to make noise individually. The smallest fingers clapping together quietly represented the agitation of a few, and the mighty noise represented the power of many. He led the crowd in a chant: "*¡Sí se puede! ¡Sí se puede!* [Yes, we can! Yes, we can!]," which referred simultaneously to the goal of putting a new station on the air and to the goal of unity and resistance more broadly (see figures 2.6 and 2.7). Juan, the president of the union, then stated in his remarks that "[our labor organization's] *lucha* [struggle] is not just for farmworkers or for 'us'—it includes gays, African-Americans, women, workers, Iraqis, Palestinians, global workers, immigrants in the United States who need legality and rights.... When you [volunteers] leave [after the barnraising is over], you didn't just put up a radio station, you are [union]istas!"[83] The label "unionistas" conferred on the

Figure 2.6
A union member rallies the crowd at a barnraising in Oregon (August 2006). Courtesy J. J. Tiziou.

Figure 2.7
Union organizers raise their fists in a "Power to the People" gesture, Woodburn, Oregon, barnraising (August 2006). Volunteer photo, courtesy PRP.

barnraising's participants solidarity with the farmworkers, emphasizing that the act of building a radio station was allied with immigrants' rights and the labor movement.

The Tennessee barnraising also had a political agenda. It was held for a group of people who wanted to use the station to combat the notion that there were no Democrats, liberals, progressives, radicals, or otherwise politically left people in their area (greater Nashville). Locals wished to create an outlet for their views and provide a tool for finding and recognizing other progressives in the area. An unofficial slogan of the station was "Unbuckling the Bible Belt." The weekend's activities culminated in a plenary gathering on Sunday night, when the station made its first broadcast. Each new station for whom Prometheus conducted a barnraising made its own decision about what to do with the first on-air moments. As noted previously in this chapter, this group decided to have a dedication ceremony in which the Nashville organizers spoke (see figure 2.8). This was followed by a radio play, after which DJs and programmers for the new station signed on and introduced their shows.

Figure 2.8
The jubilant, emotional opening moments of the Radio Free Nashville on-air dedication (April 2005). Courtesy Pablo Tao Virgo.

In the play, the people of Nashville's brains were consumed by mainstream media outlets (an unsubtle allegory). They were turned into zombies and later rescued and turned back into sentient humans by Radio Free Nashville. After the play, the programmers got on the air and people walked around listening, hugging, congratulating each other, and still crying a bit. Idiosyncratically, someone broadcasted a recording of Noam Chomsky reading his political writings over Wagner's "Ride of the Valkyries." Then Pink Floyd. Then Joy Division, "Transmission." Some volunteers drove home, calling in to report how far away they were still receiving the station's signal. Other people milled around talking, drinking beer and whiskey, and continuing to work or clean up. Elation and exhaustion were equally palpable.

The Tennessee barnraising experience seemed to have elements of a religious revival and a charismatic cult (or at least how I might imagine these experiences could be, never having actually experienced either). The event lingered with me for some time, and I know other people felt the same way. The experience of the first broadcast on this station also reminded me of Victor Turner's concept of communitas, in which social differences are leveled in a communal experience.[84] I do not want to to romanticize this experience, or to downplay the conflicts and tensions that occurred at various points over the weekend. But there was something singular and special about it. Likewise, the Oregon barnraising felt galvanizing. It truly was exciting to have Oregon's new radio station framed as a tool for political organizing by such an experienced union. It was also moving to hear their group and station stand in solidarity with other movements. Most of the weekend's activities were set up to be bilingual.[85] This fostered a sense that the event was being conducted in a way that emphasized different people coming together around a project; at work stations and meals, people interacted in English, Spanish, and halting versions of each. By contrast, the barnraising in Massachusetts in 2005 seemed a flatter experience. That group's first broadcast included a Pacifica recording about Hiroshima—a heavy subject, but a canned recording.

Radio activism in the form of barnraisings and Geek Group was a way for participants to enact beliefs about community self-determination, political participation, and freedom of expression. The application of radio technology was seen as essential to achieve these goals. As Emily, the Tennessee station's founder, said, "We're looking forward to unbuckling these Bible belt airwaves with our new station by giving real people the tools to find their voice and use it in our democracy. Plus it will be a hell of a lot of fun!"[86] These beliefs were bound up in the notion that new radio stations in the hands of "the people" will engender social change. But media reform alone was by no means the end goal of the activism, as Juan's and Rose's remarks indicate. Activists saw work with radio technology as a direct application of critical agency. They believed the propagation of radio would enable community members to create their own media and effect local and global social change.

Conclusion: Conflicts and Complementarities of Identities

Countercultural, activist, and geeky conceptions of self each figured prominently in how the activists experienced relationships with technology, with one another, and with the volunteers and community members alongside whom they worked. These identities were more than simply roles that people tried on. Rather, they functioned as resources that activists used to formulate meaning around activism and technical work. They were also used to enroll others into activism and technical engagement. Consequences flowed from the construction and deployment of these identity constructions. Sometimes these conceptions of selfhood were complementary to each other (such as geekiness and activism being performed together in the case of pirate radio or a Bastille Day celebration). But in other instances, identity constructions around radio seemed contradictory or presented challenges for the actors.

The question of technical expertise presented a real challenge for the radio activists. Indeed, there may be an essential tension between activist selves and geek or technical selves, because some of what a geek self is is an expert. These activists were dedicated to a participatory, egalitarian vision, and their iteration of geek identity was a uniquely participatory one (in contrast with, for example, many hackers). As Brian said, "Balancing making sure that stuff's working at the end of the [barnraising] weekend with making sure everyone feels a sense of ownership and has learned something is a constant struggle."[87] The barnraising is thus a singular and puzzling spectacle. Its foregrounded goals of sharing knowledge, teaching, leveling expertise, and providing hands-on participation to everyone to some degree exist in opposition to its other ceremonial purpose: to put a new, working radio station on the air. Although activist selfhood and geek selfhood each had a role to play in deepening technical engagement, tying the two together could not always be achieved seamlessly.

Activist identities and countercultural identities may conflict (or at least not coexist as comfortably as one might imagine). The New Communalist attitude toward social change tended to emphasize personal transformation as an indirect route toward broader societal shifts; New Communalists were less interested in pursuing the direct action strategies for social change favored by the New Left.[88] In the early twenty-first century, these radio activists ran up against similar issues. They designated radio technology as "appropriate" for a number of the same reasons that New Communalists and Appropriate Technology movement people did, but the radio activists also favored direct-action strategies and engagement with formal politics. They viewed access to media and freedom of expression as causes important enough for people to go to jail over them; they also toiled in policy advocacy.[89] They held barnraisings only with groups that had political missions that they viewed as vitally important. On a symbolic level, this elevated the occasion of the founding of a new station from the inchoate notion that radio is a tool for personal transformation to one in which the station was a tool for political action.

Despite sharing with the radio activists a presentation of self that was saliently countercultural or antibourgeois, some people with whom they interacted closely did not share their political fervor. Jasper once referred to a woman who was under consideration to live in his communal home as a "lifestyler."[90] Unfamiliar with this term, I asked him to explain. He said that the house members had made the decision to only admit people who worked actively for social change in some capacity or other. When they interviewed her, they designated her ineligible because she had merely the external trappings shared by some activists (superficially, piercings and a sort of punk aesthetic). Living a self-consciously "alternative" lifestyle in her consumer choices and being eager to live in a communal home were not enough, because she did not work on activist causes. This echoes the distinction between the New Communalists and the New Left. Jasper did not call her a "lifestyler" unkindly, merely descriptively. But even having this term in his vocabulary indicates that people with a more "activist" leaning may feel the need to distinguish themselves from people who are content to critique or transform society through "alternative lifestyles" or subcultural identification.

How might we evaluate radio activists' promotion of activist, countercultural, and technical selfhoods as routes to political empowerment? What are the limits of technical engagement or geek selfhood? In this chapter, I have attended to ways in which the multiple identities constructed by these actors around radio technology complemented or conflicted with one another (taking into account identity work on conscious, reflective levels and less considered ones). There is a mutual relationship between imputing a meaning to technology—radio as a tool for expression or transformation, for political empowerment, or for increasing technical or civic participation—and these meanings as social tools that guide action. As they propagated radio, activists formed relationships with radio technology that were varied, complex, and rich. Radio technology was portable, even ubiquitous. A transmitter board could fit into a lunchbox. It easily crossed boundaries of legality and geography.[91] It captivated people in basements and on rooftops, in cities and small towns. As radio technology traveled into multiple realms, activists vested this artifact with meaning in hopes of producing individual and collective transformations. Activists came to see their own selfhoods in relation to radio, and encouraged others to do the same.

3 The Tools of Gender Production

One afternoon at the Tennessee barnraising, an older electrician approached me and began to apologize for making me cry. This was puzzling to me, because I had not interacted with him at any point. I looked at him quizzically, and he quickly realized his error: he had mistaken me for another young, white woman with short dark hair. Naturally I wondered what was going on. We figured out he had thought I was a Prometheus volunteer named Louisa, and he asked me to tell her he was looking for her if I saw her. A few hours later, I bumped into Louisa, and alerted her that she was being sought. She declined to explain the situation in the moment, but later in an interview, she briefly described what had happened:

> I tried to get involved in some carpentry. And I didn't understand what the carpenter was saying, and I just ... walked away 'cause I couldn't fight the fight, I was exhausted and I was angry. I'm sure he was exhausted too. But I really wanted to be a part of the carpentry and I wanted to learn and I wanted to get involved. [Specifically,] he was talking about some kind of nail, and measuring from this point to that point, and I was kind of like, "which point again?" and he got snappish and was just like "just let me do it!" And once you start with the "just let me do its," you don't feel welcome and you don't want to be involved.[1]

I was inadvertently drawn into this barnraising conflict, which specifically hinged on issues of gender, pedagogy, novice versus expert status, and technical familiarity. Though my point of entry into this situation was unusual—I was not even involved in the misunderstanding—these sorts of disparities at the intersection of technical skill and gender were not uncommon. At the same time, such conflicts flew in the face of the radio activists' presentation of technical skill as liberating and attainable for all.

The radio activists presented the work of soldering a transmitter, tuning an antenna, and producing a news program or governing a radio station to be accessible to all. Nevertheless, they were conscious of patterned gaps in their organization and volunteer base: men were more likely than women to know how to build electronics, to be excited by tinkering, and to have the know-how to teach neophytes. This troubled the activists. They fervently hoped to provide a participatory experience that was universally attainable. The last thing they wished to do was to reproduce a hierarchy of technical participation based on gender roles. At the same time, the complex and nuanced relationships between selfhoods constructed around technical practice and gender were thorny and not easily overcome by good intentions.[2] Why did mismatches between the stated goal of egalitarian participation and actual, unequal experience keep occurring? In what ways was gender implicated?

Gendered Selves within Radio Cultures

Radio activists imagined barnraisings and Geek Group as spaces where both women and men could be empowered through gaining technical skills. Implicitly, this was held to be a radical opportunity for women. The radio activists were committed to the ideal of gender equality. They sought to realize this through the eradication of the association of technical skill with a single gender identity; they combated the notion that technical skill is equated with masculinity. As Susan Douglas (1987) and Kristen Haring (2002) have shown, a culture of masculinity grew up around radio tinkering earlier in the twentieth century. Men's masculinity and the pleasure some men attain in technical domains are mutually reinforcing projects of gender and technical construction.[3] Extending these analyses to the site of the radio activists' tinkering, it is evident that the social structure as it is in place contributes to differences in familiarity and comfort with radio hardware between women and men. The historical legacy of radio and electronics as a masculine pastime and the gendered differences in individuals' personal backgrounds with regard to technical skill contributed to a gender gap in radio activism and attendant frustrations.[4]

Masculinity takes historically and culturally specific forms. There may be multiple versions of masculinity in effect at any time. Technical competence is often a key component of masculinity. In order to maintain male dominance over new and unfamiliar kinds of machinery, men willingly adapt and modify ideas about masculinity, according to Judy Wajcman.[5] Gender categories should not be taken as monolithic. Neither traits nor competencies, for example, may be viewed as always feminine or always masculine, even in a particular moment in time. Gender is a relational system, and thus masculinity may be reinforced vis-à-vis femininity or vice versa.[6] Susan Douglas writes of the historic shift from a physically powerful, brawny masculinity to a technical masculinity in the early twentieth century.[7] Recognizing that men's relationships with technology are a dominant iteration of masculinity, the radio activists sought to challenge the association of technical competence with masculine gender. The activists were quite welcoming toward a range of gender self-presentations, spanning a range of feminine, masculine (including hegemonic and nontraditional masculinities and femininities), and queer or nonconforming. But their primary strategy was to attempt to eradicate the association of certain skills with one gender or the other.[8] As was the case in the previous chapter, the selfhoods constructed around radio became tools for social action. Technical affinity was hailed as a potential way to reconfigure established patterns of masculinity and femininity. However, activists' hopes did not always match outcomes.

A continual problem for the participants at Geek Group was that women would attend a few sessions and then stop showing up, leaving behind a small core of regular attendees who were all men. In a private conversation walking home one night, one of the men told me that he wondered if they were doing something that women found unpleasant.[9] He suggested that I as a woman researcher ought to ask the women, who, he felt, might be more open with me than with men activists. (This exchange underscores the situated nature of my account and interpretation of the gender dynamics in radio activism. My identity and appearance were also relevant in the experience with the electrician at the barnraising: what distinguished Louisa and me from other barnraising participants, but also made us interchangeable with each other at a glance, was being white women in our late 20s with cropped, dark hair.[10] His mistake was almost ironic given the activists' desire to include women.) The activists were also concerned that expert men teaching novice women reinforced the notion that women were less technically competent. They longed for women with technical skills to teach others. More than one person told me that although there weren't any in their local group, women with skills were known to be "out there." These women attained a sort of mythic status, described as exceptionally "hardcore" or "kickass."[11] It was curious that the activists felt compelled to mention these skilled women; it nearly seemed like an apology for the disparity in their own group.

In an interview, Ellen (who had occasionally attended Geek Group but did not participate by the time I was observing) told me that she felt asking questions that exposed her ignorance was not encouraged. She said that she felt that novices—men or women—who might be most comfortable in the group were people who were "quiet." When presented with something they didn't understand, these people would make a note of it and look it up later. They would come armed with this new knowledge to the next Geek Group. For her part, Ellen said that "I'm a loud girl; I've always been a loud girl."[12] She found that her inclination to ask a lot of questions was at odds with the dynamic of the group, because it clashed with the quiet nature of the work. She also felt that though the group was officially a space for pedagogy, there was a competitive undercurrent that made it uncomfortable for her to show how much she did not know by asking questions. She felt that the way the group functioned was therefore particularly insidious in terms of its inadvertent inability to include women, who were already more likely to be novices than men. These women would likely become frustrated and cease attending if they did not fit into the quietly competitive dynamic forged by the men in the group. They would also be discouraged if they could not learn tangible new things by attending the group and instead had to play catch-up independently after each session. She told me that in other settings—at home and with other friends and colleagues—she learned to build transmitters and wireless equipment such as cantennas. These experiences were more enjoyable and comfortable than Geek Group had been for her. It seems plausible that even among feminist men, their culture of hardware tinkering did not succeed in the abolition of "masculine" identity displays. The "competitive" culture Ellen described served to make Geek Group feel unfamiliar or uncomfortable to many women.

Similarly, Rose (who was no longer working in radio activism but had been involved with Radio Mutiny and Goldman) said that a few years earlier, she and another woman friend built a transmitter and a limiter explicitly in hopes of starting an all-woman-run station. This never got off the ground, but they were keen to have a station ("Sugar-Free Radio") that was not reliant on men to furnish, build, or maintain hardware. (She said that they had planned to allow men to be DJs, however.) Rose reflected that her desire to have an all-woman station was not fully articulated in her mind as a critique of gender relations at Radio Mutiny. But the collective struggled with what she felt were gender-based difficulties and the fact that one male member of the group preferred to work alone on the technical aspects of the station. Women collective members repeatedly criticized him for not being inclusive and (they felt) not making the effort to communicate effectively with them about the technical aspects of the station. The women felt he hoarded his expertise. She also recalled:

> It felt really empowering to move to the technical side of radio ... being a woman and doing that.... It was like we were in a Marge Piercy novel [feminist science fiction] ... women furtively working against the system ... [I felt like] hey, I'm a girl and

I'm soldering and using flux and I have my own soldering iron and I know exactly what this limiter and this capacitor does and where to put it ... it made me really think that I can do stuff that wasn't relegated to women traditionally.[13]

It is clear from Rose's description that soldering and "learning to use flux" were acts of self-conscious gender-boundary crossing. This was especially important to her after her experience in the Mutiny collective.

Geek Group was the more routine and internal opportunity for the group to address gender and technical knowledge together. The barnraising was a more visible, public site at which the activists attempted to enact their beliefs.[14] At the barnraising, quotidian relationships with technology and other people were suspended in order to reflexively "try to show others what they are doing or have done."[15] Quite a bit of reflection occurred among the Prometheus organizers with regard to selecting appropriate people to run workshops and work tracks at barnraisings. They made a conscious effort to recruit people with the necessary technical skills and the "right" attitude about teaching and sharing expertise. This included some women with "kickass" skills who were not part of the local Philadelphia group. For each barnraising, Prometheus allocated a certain amount of money to subsidize travel for highly experienced volunteers. Even so, it was always challenging to get the "right" people to attend. The following e-mail is illuminating. It was sent by Brian (a Prometheus organizer) to the Prometheus office e-mail list. He was forwarding a message from Robin, a woman who often volunteered at barnraisings and had a high degree of engineering expertise. Even by offering to buy her a plane ticket, the Prometheans could not bring her to this barnraising:

nooooooooooooooooooooooooooo! my heart is mostly broken by my continued failed
attempts to get non-dude engineers.
-[b]
----- Forwarded message from [Redacted]> -----
From: [Redacted]>
Subject: uanble to attend
To: [Redacted]>
Date: Tue, 1 Nov 2005 15:25:20–0800
i am sorry [brian].
i can not attend.
i feel that i need to work that weekend, in order to be able to take time off around thankstaking [*sic*].
light a soldering iron for me some time during the weekend.
:[
[robin][16]

In an interview, Brian described in detail the challenges he faced in trying to make sure that barnraisings were attended by engineers who shared Prometheus's radical vision, which he felt was at odds with a more traditional engineering "culture of exclusion." He said, "Part of [the barnraising model] is really challenging the engineers themselves to make the change. Some of them really embrace it and love teaching, some of them really don't and just want to do it themselves.... That is the biggest challenge, finding people who are going to implicitly understand why it's important for people to have their hands on this technology."[17] Engineers who value mastery and problem solving at a deep level might not be best suited to teach at a very basic level. It might be difficult for them to relinquish control to novices.

Moreover, Brian was concerned about finding engineers to teach whose social identities would not lead to reproducing the association between electronics and white masculinity when they were teaching:

> I think finding nontraditional techs and engineers is really important. I have been striving for so long to find a pool of younger women engineers, technicians who are people of color and women. It's so easy to find white guys to do this stuff.... [Like] look at me, I'm a white guy, I'm one of millions of white dudes who went to engineering school, I just happen to have more of a lefty attitude than most of them.[18]

One strategy to decouple technical skill from masculinity was to place skilled women in the role of teaching novice women and men (see figure 3.1); Brian felt that his own position as a "white guy who went to engineering school" limited his ability to disassociate electronics from masculinity, as much as he wished he could. Brian stopped short of essentializing engineers who were women or people of color for having greater native inclination to teach along the lines of Prometheus's vision. But he did speculate that because of their status as "outsiders" in the field of engineering, they might have a heightened receptivity toward the goals Prometheus hoped to accomplish at a barnraising: "A lot of engineers and techs who are people of color and women basically [are critical of the] old-school engineer way, which is a culture of exclusion, the expert attitude." He did assume that the model of empowering people through learning from "nontraditional" teachers such as women or people of color was superior to them learning from "white guys" (like himself). There may have been an added benefit to women or people of color in learning from teachers with similar backgrounds. But for Brian, it was of inherent benefit for all volunteers to learn from "nontraditional" teachers, because this would aid in the disaggregation of technical expertise from particular gender or racial identities. This constituted a part of Prometheus's strategy to "break down" gendered social structure. (I further explore the politics of race and activist tinkering in chapter 7.)

Figure 3.1
Tuning the antenna to the frequency at which the station was licensed, Pasquo, Tennessee (April 2005). Note that two women have been handed the tools and are receiving instruction from a male engineer (just outside the frame of the photo). Courtesy Pablo Tao Virgo.

Prometheus found it overwhelmingly difficult to find engineers who were expert enough themselves to solve arcane problems when they arose while embracing Prometheus's participatory vision. Brian elaborated:

> [T]here's folks like [Robin, mentioned above], she's fantastic, a perfect example of the type of people I want running the tech side of things at a barnraising, she's a good teacher, she isn't as stand-offish, exclusive, the approachability is much better with her than a lot of the other engineers....
>
> [But i]t's tough because we have to balance people who can get shit done [possess the greatest expertise] with people who can teach, and ultimately there are a couple of things that are beyond me [technically] and there are a couple of other [more expert] people I need to rely on.... It's a struggle to find people who can do everything. It's really hard [to find women].[19]

It was a constant, uphill challenge for the activists to find good teachers who would not put off novices. The dissociation of virtuosic technical expertise from masculinity in particular remained elusive.

John Dougherty, a volunteer who praised the barnraising model for its explicit desire to teach and share skills in a Prometheus newsletter, also criticized the Tennessee barnraising he had attended for what he called its "dudecore"[20] tendencies. He noted that "the volunteers just seemed to fall back on what they were used to—men doing carpentry and computer work, women doing organizing, logistics, and support."[21] Prometheus responded to this in the newsletter, stating, "Try as we might, [we] obviously [have] varying degrees of success reigning [*sic*] in the dudecore tendencies at our technical projects."[22] They solicited feedback to the volunteer's critique from the readers, and said that they would include the best responses in an upcoming newsletter. As their "anthropologist" (as they often called me), I was keen to see these responses and requested that they forward me the ones they received. As far as I know none ever arrived.

John Dougherty's first barnraising in Nashville was my first barnraising. I also had been struck by the gendered division of labor that was apparent, especially because it stood in contrast to the barnraisings as they were generally advertised by the activists. When I discussed this with Prometheus organizers, they readily acknowledged that the Nashville event had not been their finest effort in terms of distributing work equitably or in terms of having a range of experts to teach volunteers. They said that other events were more successful in this regard. Over time I noted greater variability in the distribution of skill and gender; a barnraising in Oregon the next year did have quite a number of expert women doing studio, transmitter, and antenna work (see figure 3.2). These tasks had mostly been accomplished by men in Nashville.

Figure 3.2
Three women setting up a studio. Woodburn, Oregon (August 2006). Courtesy J. J. Tiziou.

Other volunteers also shared their impressions of the work and teaching structures at barnraisings, which bring to light the complexity of the relationships between gender and the technical work the activists encountered. Clara, a thirty-one-year-old Prometheus volunteer whose day job was in computer networking, said,

> I've been sensitive to Prometheus's sensitivity [about gender and technical work]. [Brian] makes no bones about saying "It's a dudefest. We need more women." [But] I feel sheepish that I'm not cooking [because I can do both]. I feel an odd responsibility. I definitely feel like I'm extra visible and like I should be extra visible. I understand what I have to contribute by virtue of the fact that I have a nice pair of tits ... [*laughter*] ... Chicks with Ethernet cables, there's a certain inherent value in that, even in just seeing that, especially for people who aren't used to seeing it.... I'm willing to go out on a limb because it's needed, but I feel kind of self-conscious about my skill level in comparison to my visibility.[23]

Clara's technical skills made her feel like she was on display in the barnraising setting. Whereas her gender alone ("nice pair of tits") was not remarkable, when coupled with technical expertise ("chicks with Ethernet cables" embodying the two together) it became noteworthy. She reinforced Brian's statement that there was an "inherent value" in women using and teaching electronics. But at the same time, she said that she felt like she was perhaps not doing her part in the kitchen, where she also had skills and where, she recognized, a lot of work also needed to occur at barnraisings. She was comfortable with her technical skills being on display, but she was self-conscious about it at the same time.

Louisa (whose conflict with the electrician was described at the beginning of this chapter) had a background in fashion and small-business management. Unlike Clara, she did not have familiarity with electronics or software. Louisa discussed the dynamics she experienced while volunteering at barnraisings, with explicit reference to the gendered division of labor she observed (the Tennessee event was her third or fourth barnraising):

> I don't have any skills specific to building radio stations. [So] I started working the registration desk.... There's so much going on, [and] no one really takes care of the logistics part of having so many people in an area. I've waitressed, and you can kind of see how [organizational] things need to happen. There's a lot of women's work that I do.... It's hard to teach men how to think in terms of taking care of others, physical [and logistical] needs.[24]

Louisa also said that much of this work seemed invisible and was therefore underappreciated by participants: "Any time you wind up with a physical product, that job is more valued than a job that is daily maintenance. The team that is there to build the console in the studio, they wind up with a *thing* in the end, but the person that is organizing the tools, you don't have people say, thanks, great job, thanks for not losing my tools! It's just kind of expected of the job. Or the person that's washing the dishes [also would be invisible]."[25]

Clara and Louisa raised the issue of the vast logistical and organizational labor at barnraisings that was quite apart from the hardware of the radio station. Louisa accurately characterized the chaotic atmosphere that sometimes arose at barnraisings when she said that "no one really takes care of the logistics of having so many people in an area." In spite of the vast amount of advance planning that organizers put into barnraisings, they often felt very ad-hoc. During the planning and especially the execution, there was a pervasive sense that there was not enough time to accomplish the necessary tasks at barnraisings. Prometheus organizers, interns, and volunteers were often in high gear (ranging from a stepped-up and frantic pace to near-crisis mode). John Dougherty and Brian noted that there was not a great deal of formal structure to promote the integration of "nonspecialists in the building and tech elements"[26] of the barnraising. Louisa observed that formal structure to integrate participants into nontechnical elements was lacking as well. In an effort to delegate, there were signup sheets and organizers responsible for accomplishing a range of tasks. But sometimes these formalized efforts were ignored in the chaos of the barnraising. At least as much work happened due to people presenting themselves and asking what they could assist with. When people needed assistance cooking a meal or raising a tower, they verbalized this need and conscripted volunteers. The necessary work did get accomplished, usually by whoever stepped up to do it. In practice, this meant that much of the nontechnical work was done by women and technically unskilled men.

Volunteers described a variety of barriers to gender equity in barnraising labor. Louisa pointed out the fact that at the Tennessee barnraising, much of the cooking and organizational work done by women. She astutely noted that some of these women did not share Prometheus's vision for either expertise sharing or gender equity: "there were some very big domestic divisions [at that barnraising]. There were some women who were not willing to give up their domestic roles because they felt like they had power in that situation."[27] One example of kitchen people invoking power occurred when a conflict erupted between the people managing the kitchen and some "freegans" (people who do not eat animal products unless they will otherwise go to waste) who had picked up a roadkill raccoon, which they intended to eat later and wanted to store in the freezer (they were ultimately not permitted to use the freezer). She speculated that this was perhaps due to the fact that the barnraising was in the South, where she thought that more traditional gender roles might be more salient, even among progressive people. Clara also noted the reluctance of some women to expand beyond a traditional feminine domain or attempt to synthesize "feminine" skills with "masculine" ones:

> [Sometimes I'll] walk into a [place with] really cool women in one room, [and] lots of really cool men, in another room. Women, knitting or cooking, some neo-traditional urban hipster sort of activity. I like to knit, but some women seem sort of icy to me when I want to go into the other room and fuck around with a fuzzbox [an audio processor]. I feel as though they feel that I feel, you know, I'm saying I've had enough of knitting. I feel like a little bit of a traitor and like I'm leaving behind the women, to go do boy stuff....
>
> I feel like there are [lots of] examples of this: I'm talking to one friend about [an] audio driver, and then another about the baby and how she's feeling, and often they're partners with each other ... sometimes I wonder about why I can't ask the men about their feelings and the women about audio drivers.[28]

Clara and Louisa voiced discomfort with the efforts by some women who considered themselves to be politically "progressive" to define their power and their solidarity with each other through a reinscription of neotraditional gender roles, which equate femininity with domesticity. According to Fred Turner, some commune dwellers in the 1970s "did not so much leave suburban gender relations behind as recreate them within a frontier fantasy: 'A girl just becomes so ... so *womanly* when she's doing something like baking her own bread in a wood stove. I can't explain it. It just turns me on.'"[29] It is striking that Turner could be describing the situations noticed by Clara and Louisa; they each pointed to a conflict between a feminist vision of women's equality that hinged on shared and equitable work for men and women, versus a lingering New Communalist ideal that valued separate spheres for men's and women's work. Clara noted that she felt like some women implicitly challenged her status as a woman, making her feel like a "traitor" to femininity and women's solidarity. The politics and aesthetics surrounding normatively gendered activities like knitting or electronics tinkering made some people committed to gender equality feel like they were in a no-win situation.

It proved quite difficult to effectively divorce the performance of masculine gender from technical skill at barnraisings. Revisiting Louisa's experience with carpentry, she said she had sought out the carpentry work after finally being relieved from hours on her hands and knees, keeping people from tracking mud (a constant feature of that barnraising, because much of it was held outdoors, and it had rained for two straight days) into an indoor, carpeted space. She was interested in carpentry specifically because it was unfamiliar to her and because she felt exhausted from having taken on so much logistical "women's work" (in her description). She wanted to try something fresh. Her experience with the electrician was reminiscent of the difficulty activist Brian described in reining in the tendency of some people who already knew how to do a certain kind of work from monopolizing it, which discouraged neophytes from learning.

Masculine displays of technical mastery were not restricted to men; they could also be exhibited by women. Louisa described a situation at a different barnraising where female technical volunteers were present. In this case, the resistance she encountered to participating as a novice came from these women. She said, "I tried to go in and be a part of the construction team when I saw there was another woman building the studio. And I got shoved out of the way. This is [supposed to be] community [building] and I'm not here for that, so again I walked away."[30] Here one can detect a note of pain for Louisa; her experiences plugging into the technical side of work at barnraisings had been almost universally frustrating for her. Reflecting on her experience with the women, who were younger than she (though she was only twenty-nine), she conjectured, "when [some] girls turn about eighteen, they are little superstars. And they aren't quite adults yet, and they don't quite identify with other women." In Louisa's experience, there was a "macho" component to these women's display of gender and technical prowess; they chose to identify with men or perhaps with other virtuosic women, but to ignore Louisa, who was older and did not have the carpentry or electronics skills they possessed. Louisa speculated that these younger women had not (yet?) developed a sense of solidarity with other women and that this contributed to their willingness to ignore her. Performing "masculine" traits such as being competitive and demonstrating mastery ("showing off") was consciously or unconsciously opting to not decouple the skills from a gender identity. It may also be the case that acting in a masculine manner would garner recognition and appreciation from more experienced (often older, often men) techies and mark one's status as having elite knowledge.

Further complexity was added to the gender dynamics at barnraisings by the salient contingent of attendees who displayed nonbinary gender characteristics. Some of the women, particularly those found doing more "masculine" work tasks such as carpentry, antenna tuning activities, or studio wiring and set-up, enacted a self-presentation that was counter to mainstream femininity. They wore their hair short, and eschewed makeup and feminine jewelry. They did not remove body or facial hair, and wore unisex clothes. There was a high awareness and acceptance of queer politics and lifestyles in the activist group generally. They avoided heteronormative speech, for example. In some cases, relationships with technologies may be sites in which people enact or change gender identities.[31] My argument here is more speculative than bolstered by extensive evidence. But the self-presentation of some of the barnraising participants suggests that their performance of technical mastery may have enhanced a gender identity that is in other ways challenging a traditional feminine presentation of self.

Clara offered some very provocative insights on this topic. In her experience, often the people who "built the strongest bridges" with regard to technical skill and gender equality were people who were nontraditionally gendered. She spoke of cases of transgender people she knew who had transitioned from being women to being men. She said that although she was very happy to see these people transition into the people they wanted to be, it was also "sad to lose awesome women [with great tech skills and a strong awareness of gender politics] to [becoming] guys."[32]

Clara also said that in her opinion, the main reason for the lack of gender equity among people with technical skills was not sexism or exclusionary behavior on the part of the men (at least among the core group of radio activists, as opposed to external volunteers). Rather, she perceived a shortage of women willing to participate in these activities or learn and display technical skills. "The framework is there, ready for women like me to jump in—this community is ready for women with good tech skills, and they are who I most get along with."[33] Similar to Brian, she felt that part of the problem was a shortage of women who had the range of competence and comfort with radio technology. It was difficult to find women who saw technical skill as desirable and commensurate with a feminine identity. She felt that the feminist men's attempts to be inclusive toward women were genuine and mostly adequate. As a woman who was not intimidated by technology and who already possessed a high degree of technical expertise, she felt different than Louisa or Ellen, who found masculine displays during the learning process to be off-putting. In many cases, women who were most comfortable taking on the challenges of learning technical skills and exhibiting geek or technical identity were people who were already comfortable rejecting or reinventing mainstream femininity.

Conclusion: The Work of Gender (Working on Gender)

The radio activists valued craft and production at the community scale. They also strongly believed in the emancipatory potential of tinkering and an implicit radical promise in teaching women to solder. But they continually grappled with iterations of masculinity and femininity that surrounded technical and nontechnical work, some of which they wished to retool or reject outright. The numerous examples of the conflicts between novice participants and more expert people showed that people with expertise sometimes struggled to treat neophytes as equals. Novices with activist values had difficulty asserting their need or right to participate when faced with expert people they did not find welcoming. These issues became additionally vexing as they intersected with gender.

Countercultural norms were occasionally in conflict with the activists' vision for gender identity and gender equity. As is evident in Louisa's and Clara's discussion of their experiences, people's dissent from "mainstream" values in certain regards was no guarantee of a rejection of traditional gender roles. Though they possessed differing degrees of technical (and domestic) skill, they both encountered challenges from people who did not share the radio activists' vision for gender equity. In this, the activists may have suffered from the legacy of New Communalist ideals. Historian of technology Carroll Pursell writes that the Appropriate Technology movement drew on "pride of manliness from work, skill, the ownership of tools, self-reliance, and technical competence."[34] Louisa noted that, for some people, the authority derived from traditional gender roles is appealing. Clara observed that a gendered demarcation of tasks may be viewed as a means to enact solidarity with other members of one's gender. By systematically overlooking the need to balance gender in the "nontechnical" aspects of barnraising labor, the radio activists were missing an opportunity to more fully promote their vision of gender-neutral work at barnraisings. As they focused mainly on the technical side of the work, they not only valorized the technical work (and the "things" it produced, as noted by Louisa), but they also devalued the work that was nontechnical and was rather more likely to be performed by women, as noted by Louisa, John Dougherty, and Clara (who felt that at a barnraising she could as comfortably wield Ethernet cables or cook a meal). Prometheus somewhat reined in the masculinity of the technical work by trying to decouple it from a masculine gender identity, but they were less successful in promoting a gender-neutral vision of tasks such as cooking, organizing tools, running registration, and keeping workspaces clean. In this, the activists inadvertently reinscribed the gendered associations of the nontechnical work.[35]

Conversely, activist and gendered selfhoods could, at times, reinforce one another. The feminist ideals held by the men and women in the group enabled them to occasionally dissociate skills and work tasks from gender identity. This may have also heightened the acceptance of nonbinary gender performances. Some people in the group of activists self-consciously forged gender identities that incorporated participation in tasks that would not traditionally be associated with their genders. Simon's penchant for knitting (described in chapter 2), Clara's fondness for cooking and "fucking around with a fuzzbox," or Ellen's and Rose's eventual self-taught proficiency at soldering and "using flux" were examples of this. Notably, all of these people identified as geeks.

The relationship between geek selfhood and gender warrants particular scrutiny. Ron Eglash has argued that geek identity (and the whiteness and masculinity embedded in it) can serve a gatekeeping function. This restricts nonwhites or women from embracing a geek identity. In turn, members of these "other" categories, in some cases, improvise or innovate different strategies for attaining technical access or identification (with varying degrees of success). Within radio activism, geek identity was intended to be a universally attainable identity. Geek identity may form without concomitant technical expertise, though increasing technical expertise may serve to heighten confidence and bolster technical affinity. One of the radio activists' strategies for promulgating geek identity was to decouple technical skill from gender identity, thus making technical affinity and geek identity gender-neutral propositions. Geek selfhood, although still a marker of "outsiderness" vis-à-vis the mainstream, was imbued with inclusiveness. The radio activists plainly intended for geekiness to transcend other identities people constructed and experienced. Harmony of identity was intended to smooth over disparity in skill to a significant degree: people were encouraged to identify as geeks even before they attained high degrees of technical expertise (see chapter 2).

Although the radio activists offered a version of geek identity that could harmoniously coexist with masculine, feminine, or queer gender identities, there were difficulties. Some iterations of femininity rejected geek identity as incompatible. This was the case when Clara felt the need to question whether she was a "traitor" if she left knitting women to fool around with audio equipment. Masculine identity displays in the context of technical work could serve to alienate women. In the cases of Ellen and Rose, each woman decided to continue pursuing technical knowledge anyway, but opted to do this in environments that were either women only (Rose) or simply not male dominated (Ellen). But these women were already highly committed to radio and possessed geeky conceptions of self (particularly Ellen). Louisa's experience was different. She largely retreated to what she knew. She decided she did not want to "fight" any more, so she hewed to the path of least resistance, which meant "women's work" such as logistics, organization, and keeping things clean.

Geek selfhood did not utterly lack potential as a resource. Especially as it intersected with feminism or activist identity, it seemed to offer some tools for the promotion of egalitarianism. It challenged the limits of the gender binary. Yet it did not seem elastic enough to make significant headway in terms of the inclusion of nonwhites.[36] The association of geekiness with whiteness is entrenched, even among activist and antiracist people. The core group of activists and volunteers who were active in Philadelphia and on the barnraising circuit were mainly white. (Chapter 7 engages issues of race and ethnicity more directly.)

In the end, gender cut multiple ways for the radio activists. They did enjoy some success challenging the exclusive association of technical affinity with masculinity. But their exuberance over cultivating technical affinity was insufficient to overcome the durability of gender constructions the activists encountered all around them. Their experience reminds us that just because geek and gender selfhoods are "constructed"—as opposed to innate, fixed, or immutable—this does not mean they are not also "real" in the sense of being entrenched.

4 The Work of Pedagogy in Technological Activism

At the end of a sunny weekend, activists and volunteers banded together to move two enormous refrigerator-sized FM transmitters into storage. After much hand-wringing about where the transmitters might fit, they were designated to be housed in a warehouse art space a couple of miles from the activists' office. (The warehouse was famous in activist circles as the site of a police raid on artists making street theater puppets in advance of the Philadelphia Republican National Convention protests in 2000.) The transmitters had been the centerpieces of a weekend-long tinkering workshop attended by activists, volunteers, and engineers. The ostensible goal of the weekend was to repair the transmitters. They were not in working order and had been out of use for decades. They were heavy to move, difficult to see inside, filthy inside and out, and missing various components. After two full days of diagnosing and repairing the machines, arguably little progress had occurred. The transmitters, though cleaner, were still not functional. Nor were they even significantly closer to working.

Though another transmitter work session was planned in vague terms, the transmitters never came out of storage. Their "temporary" warehousing was effectively permanent. This might seem puzzling as an episode of media activism, in that no usable hardware resulted. But the radio activists seemed unfazed and untroubled by the lack of headway. It is thus worth considering what *besides* hardware is being "produced" in episodes of technological activism. The past two chapters have argued that technologically oriented activism is a site of identity work and production. This chapter highlights the ideational and affective elements of radio activism, arguing that along with technical artifacts, the radio activists endeavored to build social relations and a politics of participatory expertise. I thus generate a sense of the radio activists' priorities in which their role as propagators of technology can be clearly heard. In order to tease out what radio activism *is* for its participants, I assess the organizational forms that the radio activists' pedagogy embodied (as well as antecedent cultural forms that flowed into their practices). I also examine the work objects and products of activism, as well as the blurring of work and leisure.

Though "activism" and "activists" are the subject of various scholarly accounts, the actual granular practices of activism are often not defined or clearly understood.[1] My attention to activism at the level of practice is based on the methodological proposition that studying media activism this way helps to illuminate *what is actually occurring in practice,* a surprisingly useful proposition. Observation is especially important to understand work. According to Stephen Barley and Gideon Kunda, "[this is] because most work practices are so contextualized that people often cannot articulate how they do what they do, unless they are in the process of doing it."[2] Julian Orr adds, "the ethnographic study of work practice ... must be done in the situation in which the work normally occurs, that is, work must be seen as situated practice in which the context is seen as part of the activity."[3] In this chapter and in chapter 5, I draw largely on observation. This enables the analysis to extend beyond taking the activists' accounts (especially post-hoc accounts) for granted. The activists' stated understandings of their efforts are also significant, but studying practice as it occurs avoids certain problems associated with "taking the actors' word for it."

Attention to practice is also consistent with examining activism as work (a theme of this and chapter 5). (Admittedly, activism and work are somewhat imprecise categories.) Thinking about radio activism as work requires the boundaries of "work" as it was understood for industrial occupations and bureaucratic firms to stretch and bend, and it is not my intent to fit activism into or argue against these categories. Rather, I bring a considered rendering of practice suggested by Orr and Barley and Kunda to the site of activism in hopes of revealing the multiple ideational and organizational forms at play in this setting.[4] Foregrounding practice changes what can be known about activism (and is of equal value in studying other sites of cultural production and mediation).

Cleaning and Diagnosing the Transmitters: Work and Pedagogy in Action

The goal for the weekend was to clean and repair two large transmitters donated to the group after a college radio station in upstate New York decommissioned them during an equipment upgrade. Participants included: the paid full-time Prometheus staff (four people); interns (the group had a rotating cast of interns, a semester, summer, or academic year at a time, usually with two to three at any given moment); novice volunteers (from

Philadelphia and New York City, most of whom had paid a nominal fee to participate in the workshop and learn about hardware); five highly skilled engineers the radio activists had enlisted to help troubleshoot and teach (some more local, some from as far away as Washington State and Illinois). I was also on hand as a participant-observer. Over most of the weekend, workshop attendance ranged from fifteen to twenty-five people. The engineers and volunteers held different sorts of day jobs, some in community media, some in unrelated fields. All of the engineers helped build community radio stations on a regular basis, though in most cases on an unpaid, voluntary basis. Most had formal engineering training of some sort. The deepest technical expertise in the group resided with them.

The two transmitters were big, in every sense. The radio activists advertised the workshop by appealing to a technological sublime: "Do you lie awake nights having technophilic fantasies about high-wattage broadcasting? ... [These] transmitters were built in the early 70s, and look like something out of *The Bionic Woman*.... [Come learn] how these behemoths tick."[5] The transmitters were forty years old and contained large vacuum tubes rather than transistors. They were not working in part because some of their original components contained PCBs that had been removed due to toxicity concerns. When operational, one transmitter produced a radio frequency (RF) signal of about 10,000 watts, and the other, around 1,000. By marked contrast, LPFM transmitters the activists commonly used could not exceed 100 watts by law (about the same power as an incandescent light bulb). Even if they worked properly, the big transmitters were unsuitable for use by an LPFM station, so the activists conjured a possibility for their use outside of the United States: "we plan to send them to Central or South America for use by community broadcasters."[6]

The transmitters were unfamiliar to the core activists because of their power and scale (as well as not being solid-state). They were about the size of refrigerators and weighed so much that they couldn't be brought into the activists' usual workspace in a church basement. They would damage the wood floors and they would be all but impossible to bring up and down stairs. Thus, storage and moving concerns were nontrivial. The activists rented a large truck and a pallet jack in order to move the transmitters to the workshop site. They conducted the workshops inside the truck bed and outside on folding chairs and tables set up on the church sidewalk next to the truck (see figure 4.1). The transmitters also required serious electrical current to run, which meant trying to secure a generator. This proved expensive and difficult. Ultimately the activists decided that

Figure 4.1
Volunteers unloading a big transmitter from a truck. Author photo.

the transmitters needed so much work that the workshop would be conducted without powering them on, which would save the cost of renting a generator.

The workshop was organized to resemble a "minibarnraising." This meant that explicit teaching tracks ran alongside ongoing work on the transmitters. People moved fluidly to drop in and out of formal and informal activities, taking breaks for meals with the whole group. The engineers focused on cleaning, assessing, and diagnosing the hardware, and the staff activists ran lectures and tutorials for the novices. There was much drift as volunteers and staff idly moved from the formal workshops into the trucks in order to clean, ask questions, or observe. The engineers would sometimes work on removable parts outside at tables in full view, and make attempts at explaining what they were doing. The workshop was outside in a public place and the weather was pleasant, so the Prometheus organizers set up a table with their literature. They hoped to chat with passersby about the workshop and the organization generally (and possibly solicit donations). The truck was festooned with a hand-written sign that read, "What are these crazy people doing inside that ginormous truck? Come in and find out!" This improvised attempt to garner publicity represented the activists' goal of expanding participation. Although it was not necessarily a realistic goal, they would have eagerly welcomed the addition of neophytes off the street (see figure 4.2).

On the first day, Jasper gave a talk about the technical properties of radio, providing an overview of RF, electromagnetism, and hardware, addressed to the novices in the group. Simultaneously, people worked to clean the transmitters with rags and toothbrushes. They performed diagnostic work, which was dominated by the most experienced engineers. Early on, the engineers determined that the more powerful transmitter was in better condition than the 1,000-watt one, so effort was focused on the 10,000-watt machine. Jasper himself had a deeper engineering background than the volunteer attendees, but was largely self-taught. He was less knowledgeable

Figure 4.2
Soliciting passersby from the truck. Author photo.

than the engineers brought in for the event. He was also less experienced than Brian, the only staff activist with formal engineering training. Jasper's lecture included an introduction to the parts of a radio station, antennas and standing wave ratio, electronics components found in a transmitter, and power, moving between political and technical registers (and even punning to connect them). He explained resistors:

> This is a good word for radicals who are against the state. [*laughter*] It's measured in ohms. Think about water. The bigger the tube, the less resistance it encounters as it goes through the tube. Resistance is not in itself a bad thing. Sometimes in a circuit there are advantages to not letting all the power flow. A light bulb is a resistor. It makes electricity flow slowly and heat up the filament and turn it into light.[7]

In tying political radicalism to ohms, Jasper connects a political stance to technical details in keeping with the activists' wider political project. Jasper also displays the activists' idealized model of expertise, stressing that recently he too had been a novice and taking pains to promote egalitarianism in technical practice: "One of the good things about me teaching you is that I don't really know that much about radio. I'm not that far ahead of you, as opposed to people who know way more and are basically incomprehensible."[8] The activists were not only engaged materially with the radio transmitters. They were engaged in mediation of these artifacts, casting even the inner workings of electronics with a valence of radically egalitarian political participation.

The equipment was very dirty. Nearly everyone took a turn over the course of the two days scrubbing inside the cabinets that held the components. Delicate or particularly dirty pieces were removed for special cleaning (see figure 4.3). A silver-plated vacuum tube had to be dusted and then polished. It came out nicely. The most important diagnostic task was to see if the exciter worked. The exciter is the part of the transmitter that actually produces the RF. It could be assessed with tools and instruments running on ordinary house current because it only needed to put out around 300 watts. Other components would serve to amplify this signal from the exciter to 10,000 watts, but without a generator there was not enough current to power up the transmitter. (And most participants were relieved by this. One of the engineers, Jim, told everyone: "These big transmitters are dangerous. They must be used with respect. No one should ever repair, maintain, or even turn them on alone. They are deadly and you need another person to push you away if you start to fry!"[9]) Assessing the condition of the exciter was a task largely dominated by the engineers, and novices did not participate, other than to hover around them and ask a few questions. At the end

Figure 4.3
Volunteer cleaning inside the cabinet while engineer looks on. Author photo.

of the first day, Jasper and Brian asked the engineers to summarize the work that had occurred, including the diagnostics performed on the exciter.

The second day of the workshop was much less structured than the first. There were no lectures or formalized activities, but cleaning, testing, and tinkering with equipment all continued. Novice participants essentially found themselves restricted to either cleaning or helping with communal meals. None too pleased with this division of labor, they took breaks in which they sat around chatting with one another and ruing the fact that they did not know how to "plug into" the technical work. They complained that they did not feel especially welcome to do so.[10]

In practice, the activists' desire to propagate egalitarian technical participation was difficult to implement. Jasper had tried much harder than the engineers to make himself seem non-expert and accessible (for example in his lecture). Reflecting later, he was critical of himself and of Prometheus for not having tried harder to implement the oft-repeated barnraising ideal of "no one being allowed to do anything s/he already knows how to do" over the weekend. The big transmitter workshop was a special "one-off" event in some regards. Far less planning had gone into it than went into actual barnraising events. The experience of the weekend amply demonstrated that without aggressive measures to combat the hoarding of expert knowledge (deliberate or not), Prometheus's vision for ecumenical skill sharing could not be realized.

Jasper and Brian experienced a special tension in this situation. They had much more technical knowledge than the novice participants but they had less expertise than the visiting engineers. As a consequence, they were torn between trying to learn more themselves (and extend their own understanding of the technical problems) versus trying to make sure the engineers included the novices in the activities. They wanted to do both, but these goals were at odds with each other. Jasper and Brian repeatedly stopped the engineers to ask them to explain what they were doing while they were doing it. They also insisted on an accessible and public work report at the end of the day. Brian, in particular, had a very gentle yet persistent manner. He would not permit the engineers to brush off his inquiries, repeatedly urging them explain their activities to the group.

Nonetheless, a dynamic familiar from preceding chapters emerged. The group contained very expert members and many tasks needed to get accomplished. Giving the novices a full and comprehensive understanding of the work that was taking place would have impeded the engineers' ability to learn as much as they could about what was wrong with the transmitter. The engineers were not terribly interested in slowing down to explain, let alone giving over the equipment and diagnostic tools to novices. Similarly, when the novices paused in their cleaning activities to learn more from the engineers, it slowed the massive cleaning undertaking that was underway. As a consequence, novices mostly stuck to what they knew they could do and were disinclined to cease being "productive" or interrupt the engineers. Hence technical novices primarily cleaned and provided meals over the course of the workshop weekend, and technical experts primarily performed the tasks that required technical expertise. At the end of the weekend, the transmitters were not repaired and they needed to go back into storage (the exciter worked, but on its own it was unimportant). Vague plans were made to bring the engineers back to have another go at the project, probably with a generator. Notably, this was not planned as another pedagogical workshop. It was to be a more expert group focused on getting the transmitters running. This decision represented an acknowledgment that supporting egalitarian technical practice was an uphill battle. It was also an acknowledgment of the failure of the group to fully implement their ideals in this case.

Interpreting the Work of Pedagogy

I next turn to addressing a series of questions raised by the workshop: How does the activity surrounding these broken radio transmitters provide an opportunity to think about the underlying dynamics, objects, and motivations of radio activism? What were its objects of focus? What factors motivated the participants in radio activism? And what were the precedents for and implications of how the workshop was organized? How did the disparate activities of the workshop coalesce into a coherent and meaningful experience for the activists? Lastly, what are the contradictions between the objects of focus and the forms of practice revealed in the activists' practice, and why do they matter?

Objects of Focus

Different people held different goals for the weekend in the transmitter workshop. The engineers most desired to troubleshoot and fix the transmitters. Novices wanted to gain familiarity with electronics and thus acquire a sense of agency over technology. The activists' goals were perhaps more varied and subtle. However, it may be too simple to merely state that activists', engineers', and novices' goals varied. Meaning was being made (and work practices were being focused) around more than one thing.

Thinking about focal point (or what sociologist Monica Casper calls the "work object"[11]) for the media activists illuminates aspects of their practice that would otherwise be hidden or difficult to interpret. Of course, the goal of the weekend was ostensibly to clean, diagnose, and repair the big transmitters. This might seem to indicate that the work object (and ostensible work product) is hardware. Radio stations are the most obvious products of barnraising events, after all. And staff activist Ellen's e-mail signature read, "building radio stations = awesome."[12] This e-mail signature can be read as a performance indicating that this product was of great significance or even primary importance. This was certainly the case for the engineers. They were largely absorbed in trying to get the equipment to function and were sometimes frustrated at all of the cleaning and other work that had to be accomplished before they could immerse themselves in the diagnostic and repair work. For them, working hardware was a very significant object of focus, if not their sole one.

Yet to understand the work object in these efforts as simply radio hardware is to take an unduly narrow view of the radio activists' project. In this workshop and other technical activities, they sought to "produce" widely distributed technical expertise and a politics of engagement; the ultimate goal of these activities was desirable (i.e., "democratic" and "participatory") social relations. At the transmitter workshop, social relations were arguably a more significant product than radio transmitters, because at the end of the weekend, it was not remotely troubling to the activists that the transmitters were nowhere near producing an RF signal. This underscores the point that different people engaged in a common work project may not all share an understanding of what the work object is. The lack of a straightforwardly identifiable work object in this setting provides an opportunity to interrogate overlapping but not identical conceptions of activist production.

The radio activists' focus on social relations including participatory expertise was evident at many turns during the workshop. Staff activist Brian said, "A lot of the old-school dude engineers, they don't always get it. Or they get it, but they don't know how to put it into practice.... Like [Jim, who] is a prime example of a not-approachable engineer, he's a fucking grump, and if he weren't such a genius I don't think we would want him there."[13] The engineers were not always adept at balancing the need to produce hardware with the need to produce egalitarian social relations, even if they "got" Prometheus's vision. (Notably, what Jim said about the perils of the big transmitters—"you will need someone to push you away if you start to fry!"—was potentially intimidating to novices.) Yet Prometheus could not dispense with the engineers entirely if they wanted the machines to function. Certainly novices and even staff activists did not possess deep enough knowledge of electronics to be assured of fixing problems by themselves. This tension between the work object as hardware versus social relations was also exhibited in the conflict felt by Brian and Jasper who felt pulled between using the weekend to learn from the engineers and thus enhance their own technical skills, versus concentrating on making sure that novices were included and enculturated.

Activists' efforts were not primarily focused on producing technical knowledge. In an interview, staff activist Thomas that building radio stations was actually "just sleight of hand" for their "real" work, which he characterized as grassroots community organizing with a radical left agenda. He felt that this agenda was not limited to media issues; he broadly characterized their goal as "community empowerment" and critiquing dominant power relations.[14] The orientation for radio activism in Thomas' understanding was decidedly outward and extralocal. He perceived Prometheus's work as supporting a global social movement.[15] This echoes the activists' repeated claim that "a barnraising isn't the most efficient way to build a radio station, but it is the best way to build a movement."[16]

The activists routinely declared that "community radio is 90% community and only 10% radio."[17] Crucially, they attributed this quote to Zane Ibrahim, a broadcaster on Bush Radio, a community station famed for challenging the apartheid regime in South Africa. This again symbolically linked the activists' local practices to the idea of a global movement and underscores that this ideation was a significant aspect of their work. Propagation of not only artifact but prescriptive politics was the heart of their project.

Though it is clear from these examples that the radio activists thought and spoke of their work objects and products in more than one way, it is unhelpful to insist on a strict analytical demarcation between hardware as work object and politics or social relations as work object. Anthropologist Christopher Kelty argues that free software geeks imagine their social existence through technical practices as much as through discursive argument. This idea resonates with radio activism. Even when focused on a nominally technical project the radio activists were never exclusively addressing technical concerns; they were also addressing and enrolling others through their technical practice.[18] Thus, technical practice served as site for training and refinement in two domains highly prized by the activists: technical practice and desirable social relations. The affective pleasure felt by activists in tinkering with hardware and affirming a politics served to strengthen the connection between these domains for them. They also wished to shore up this connection for workshop participants, though this required effort.

Motivations: Leisure, Occupation, Remuneration, Gratification?

If the weekend's objects of focus varied among participants, so did their motivations for participation. For staff activists and interns, this workshop was part of their jobs. Yet the workshop was held over a weekend: thus staff, interns, and outside volunteers (expert and nonexpert) "donated" "free" time to the organization (and to their own affective and educational priorities).[19] In certain ways, this made sense given that ham radio (LPFM's precursor) was predominantly a leisure-time activity. Barnraisings and the weekly Geek Group meeting (held on weekday evenings) conformed to this pattern as well. Yet of course, activism is not defined by strict demarcations between work and leisure or between work and volunteerism.

In her study of radio hams in the United States from the 1940s to the 1970s, historian Kristen Haring shows how the hams' leisure electronics practice had implications for their paid occupations. Hams often were employed in technical fields; thus ham radio conducted as a hobby also served as occupational training.[20] Similarly, some of the expert engineers Prometheus invited to the workshop were also employed professionally as engineers. For them, puzzling through broken electronics could be viewed as relevant to their skill set for their paid employment. Work has largely come to be defined not merely as a productive effort, but as a social relationship having to do with capitalistic exchanges and activities defined and described in the terms of employment.[21] Yet understanding activism as wage labor paints an incomplete picture. The core group of activists and interns spent their weekend at this workshop without an expectation of additional remuneration for their participation. These hours were recuperable as furthering activist goals, not because anyone was paid for them.

The category of "amateur" may offer a fruitful way to look at the intersections of work, leisure, and volunteerism. "Amateur" in its etymological origin ("lover") connotes something closer to a vocation. Being an amateur may carry a stronger connotative weight than a mere pastime. A strong sense of affective connection or affinity is cultivated alongside the material practice of amateurism in any given domain.[22] For radio activists, tinkering and "off-the-clock" volunteer pursuits may also be considered a form of "occupational training." Devotion to these activities was consonant with their paid work (and may even have built up skills they used in the workplace). Activists routinely engaged in activities outside of their actual work setting that seamlessly extended into their work for Prometheus. However, electronics tinkering as conducted by these radio activists was arguably less important for the training it provided in electronics. It was a more significant site for the media activists to construct a technical affinity and to imbue technical practice with political significance. Through these practices, they shaped the meaning electronics tinkering as being part and parcel of activism (not merely a hobby, which it also resembled). This also had implications for selfhood: tinkering here served to cohere worker identity and personal identity more generally. "Leisure" time was often spent furthering the same goals as "work." In analyzing these activities, it is important to pay attention to how participants felt and what they experienced, not only what they produced.[23] Tinkering and technical participation were enjoyable for many members of the activist group, as well as for the engineers.

Staff activist Thomas said in an interview, "We would be doing this anyway, we're just lucky to be getting paid for it."[24] This indicates that work identity and a wider sense of self were intertwined. Identity as a worker was not segmented from social identity more generally. The group relied on paid as well as unpaid and volunteer labor in order to maintain its productive output. Unsurprisingly, the paid staff members always worked many more hours than they were paid to work. During this period, the full-time staff members were paid for forty hours of work per week at $10 per hour, though all of them worked well over forty hours. In leaner times, they were paid at twenty and thirty hours per week based on what the organization could afford. They occasionally laid themselves off and collected unemployment when grant money ran out.

Employees were asked to perform electronics diagnostics, give lectures, and cook meals, as well as file papers, write newsletters, go on speaking tours, clean the office, maintain spreadsheets, ally with other groups on campaigns around media issues, and enter the halls of power in Washington, DC (all of which are discussed in chapter 5). The definition of work activities as formalized tasks made explicit in a contract between employee and employer hardly applied at Prometheus. A description of all that an activist job encompasses is always incomplete. Julian Orr describes this gap between formalized expectation and practice: "a possible conflict [exists] between work as doing, as practice, and work as activities explicitly described."[25]

Prometheus workers conceived of their work as *productive* activity (as opposed to *paid* activity). On the surface, Thomas's claim that the activists would be engaged in their activities with or without getting paid for them sounds implausible. Real-world concerns—rent, food, health care—simply must intervene on some level. Yet two of the four paid full-time workers began as volunteers for Prometheus (without expectation to transition to paid employment with the organization, given its scale and fragility in the early years). None of the activists had class backgrounds that enabled them to freely choose unpaid work without concern for money. They all chose to live very modestly, which was possible in Philadelphia (a relatively economically depressed postindustrial city) during the 1990s to 2000s. Surprisingly, staff members even managed to own homes (or lived with partners who owned homes). This was one potential factor that provided stability and fixed rent payments, enabling them to choose low-paying work. Clearly, volunteering without being supported by a partner (or other independent means) would not be a viable proposition over a long period of time.

This raises the question of what is gratifying about activism, if not remuneration. Certainly, the ability to tinker, solve problems, and play with machines is fun for some people.[26] Additionally, the radio activists imagined many points of connection between their activities in the transmitter workshop and other iterations of media activism and community broadcasting. One instance of this was the possible future they articulated for the transmitters: the activists and engineers claimed the transmitters might find their way to Central America, where community or pirate stations could use them. Even (or especially) when the work itself was boring, dirty, or frustrating—as it was when the group failed to repair the transmitters—activists' morale could be recuperated through the understanding that work was related to activist goals. This points to the affective and ideational work of activism. It also indicates that a significant project of activism may be the reconciliation of apparent contradiction in order to ascribe coherence and political import to a variety of activities (a theme I take up again in chapter 5).

Motivation and gratification are important to consider in terms of their relationship to modes of "production." It is routinely claimed that collective collaboration (also known as "peer production") is especially gratifying for participants.[27] However, the radio activism case shows that some of what proponents have tended to assume about peer production is less evident in practice. In this example, a nominally nonhierarchical or "network" formation did not result in a universally rewarding experience. As I show in this chapter, the work to clean, diagnose, and repair these transmitters resulted in differing experiences among participants and was far from universally gratifying. When the participants defaulted to doing what they knew how to do, the novices were relegated to cleaning the transmitters or making meals for the group, and they did not particularly enjoy this. The second day (when there were no formal activities for them) was especially tedious. By contrast, the engineers were highly gratified by the opportunity to puzzle through the transmitter's components. They performed tests using diagnostic equipment, and otherwise participated in intense and engaging work with the transmitters. For them, the least satisfying element of the weekend appeared to be the activists' insistence that they slow down work to include neophytes. They did not relish explaining their process or ceding power over the hardware in order to let novices "put their hands on it" and understand its components. At the same time, the engineers worked regularly with Prometheus and were accustomed to Prometheus's participatory vision. So it was not the case that they were opposed to including the neophytes. Mainly it was a struggle to balance their technical curiosity (and perhaps habits of engineering culture) with open participation. As a result, the engineers were somewhat at odds with the novices, even though they were not trying to intimidate them. The engineers were happiest puzzling and tinkering while talking at a fairly high level technically. The volunteers were least able to "plug in" under these circumstances, needing formal explanation and direction in order to understand, let alone contribute to the work. Novices (including myself) felt on occasion that stopping work too many times to ask questions could pose a hindrance; thus one's options were cleaning or idling. The least gratifying parts of the experience for the novices in particular were the most self-organized, least structured parts of the workshop.

The staff activists themselves had a curious role. The two with engineering expertise (Jasper and Brian) worked to increase their own technical expertise by learning from the engineers. (Notably, the staff activists' questions were not regarded as intrusive, probably in part because they had more skill than the novice volunteers. Their status as organizers of the workshop presumably also afforded them more confidence to interrupt and afforded them more potential tolerance from the engineers.[28]) Jasper and Brian also strove to expand the expertise of neophytes through the formal instruction and active participation (even when this participation impinged on the autonomy of the engineers). They imagined aloud (and repeatedly reminded the rest of the group of) a possible eventual future for the working transmitters (even though this was unlikely in reality). This ideation rhetorically justified the weekend's activities, and this was perhaps especially true for the staff activists.

Another shortcoming of what Daniel Kreiss and colleagues term the "peer production consensus" is that it masks the fact that the dynamics of peer production may vary widely depending on the site and type. Open source software projects, for example, have traditionally not focused on the technical participation of beginners,[29] resulting in very different dynamics than those experienced by the radio activists. The radio activists' case encourages an examination of the interplay between engineering expertise and an activist politics of technology devoted to "participation."[30] Yet even as it reveals heterogeneity within experiences of peer production, the radio activism workshop exhibits commonalities with facets of participatory culture. These include the group's self-organizing bent, their mentoring-pedagogical dynamics, and their cultivation of social connections among members.[31]

What motivated people to engage in the activities of the workshop cannot be explained by the supposedly inherent gratification of peer production. In fact, collective collaboration here was not gratifying; some of the participants (engineers) were merely hindered, while others (novices) were more actively frustrated. Nor was remuneration a motivating factor. Only the affective and ideational components of activist production—of radio transmitters and of utopian politics—allowed for the redemption of hard, boring, and only dubiously fruitful labors over the course of the weekend.

Dynamics and Politics of Structure and Self-Organization

The transmitter workshop contained elements of contradiction in its organization: it was highly self-organizing, though parts of it were also formally structured. Where did these opposing impulses come from? And what consequences were there for both the material and affective outcomes of the workshop? Radio activism does not directly flow from any single cultural antecedent. However, some lineages include ham radio, participatory democracy in the New Left tradition, "participatory culture" or peer production, and the Appropriate Technology movement. (Other threads of origin not discussed here include socialist anarchism, pirate radio, and DIY culture.[32]) It is beyond the scope of this chapter to fully limn these connections, but in the following section I sketch some of the most important aspects of peer production and participatory democracy as embodied in radio activism. I also revisit the legacies of ham radio and Appropriate Technology.

At least two of the engineers in this workshop had radio amateur ("ham") licenses and experience. Historically, hams possessed a close relationship with radio technology.[33] Radio activism mirrors ham culture in some ways, most obviously in terms of the constitution of community around radio technology.[34] The transmitter workshop introduced novices to radio, offering them "membership" in this activist technical community. But the workshop was oriented toward solidarity with activism outside the local gathering. The notion that other geeks and media activists were "out there" working on related goals was important as a motivating factor, and thus the workshop was not understood to be occurring in isolation. This affective and community-building function was critical. Electronics tinkering was arguably less important for the training it provided in electronics. Even after this workshop, true novices would face an uphill battle building a radio transmitter. (This was true with barnraisings as well. Although a much-vaunted feature of each barnraising was a soldering station where novices learned to solder transmitter boards, these boards were never used when new LPFM stations went on the air. Thus the new station was not dependent on novices successfully assembling this critical piece of hardware.) This demonstrates the "sleight of hand" employed by the activists, who sought to propagate not only artifacts but also a politics of technology and a community around technical practice. Tinkering's greater import was that it provided an opportunity for the radio activists to fuse technical practice with community building. Through these practices, they transformed a mere hobby into media activism.

The radio activists' explicit political heritage is complex. They were largely aligned with media reform efforts. A significant part of Prometheus's focus as an organization during the late 1990s and early 2000s was expanding LPFM through the legislative process. They also joined other campaigns and mounted lawsuits over media consolidation. Thus they had firm footing in the 1960s New Left tradition of political organizing: "the New Left may have sought to build a new world, but it did so using the traditional techniques of agonistic politics."[35] Writing specifically of the Students for a Democratic Society (SDS), Francesca Polletta argues that the exact meaning of "participatory democracy" in the New Left context was not fully clear. But in contrast to how the concept has sometimes been cast (including in the recent efflorescence of Occupy Wall Street), "it did *not* mean consensus-based and leaderless decision making"[36] at that time. Polletta describes a delicate balance in political meetings between bureaucratic, procedural norms and an effort to cultivate an ethos in which participants were treated fundamentally as equals: she writes that participants understood that certain "people were leaders before the meeting, and they were leaders after the meeting. But during the meeting they were not [leaders]."[37] Participatory democracy was intended as a means to surmount barriers of status or access to the political process, though it was not intended to subvert "structure" per se. Polletta quotes an SDS participant who recalled, "Much care was expended to encourage reticent members to express their views.... Ideas and questions were responded to without condescension or acrimony."[38] Similar to SDS, the radio activists explicitly valued egalitarianism but grappled with issues of power and structure in this workshop and in the organization more generally. In particular, power in the form of technical expertise was unequally distributed in the group. The SDS participant could have been describing how Brian and Jasper gently steered workshop dynamics to allow novices to speak and ask questions.

Yet even though Prometheus owed an obvious debt to New Left politics, their political stance was more intricate. Their activities included mainstream advocacy and coalition building around legislative goals. They adopted formal consensus in internal decision making (codified in 2007, after the period of this fieldwork). They valued but could not fully embrace piratical politics (though they made frequent reference to their pirate origins). Although they expressed deep sympathies toward pirates (including early heroes of micro-broadcasting), they accepted that in practical terms, they needed to cultivate credibility with the FCC and lawmakers (a stance I discuss in chapter 5). (One way to read the radio activists' ideation around the not-even-close-to-working big transmitters, supposedly bound for unlicensed use in Central America, was as piratical display—albeit a fairly defanged one that would in no way compromise their standing with US regulators.[39])

Peer production (or participatory culture) is also relevant here as a related mode of cultural mediation. The radio activists' workshop possessed features that made it distinct from digitally networked peer production. Namely, it was not digitally networked, distributed practice (though some elements of media activism are). It occurred face-to-face, but its contours otherwise strongly resembled some of the features scholars of peer production have named as most significant. In particular, the workshop represented nonmarket and nonproprietary collaborative practice (even though, as previously noted, its products are less than completely straightforward to identify).[40]

The radio activists inherited elements of their practice from the Appropriate Technology movement of the 1960s to 1970s. Appropriate Technology had origins in "the convergence of a broad countercultural movement, a reassertion of doubts about the role of technology in American life, and the burgeoning environmental movement," according to Carroll Pursell.[41] As noted previously, one activist claimed that "microradio was the solar power of the 1990s." This connection was strikingly apparent in the older engineers at the transmitter workshop. They were interested in carpentry and alternative energy as well as electronics and communication technologies. Ranging in age from their forties through sixties, some were old enough to have perhaps cut their teeth in the original Appropriate Technology movement. As Pursell notes, a central claim of the Appropriate Technology movement was that these technologies "worked in gentle partnership with nature and fostered intimate personal relationships."[42] This idea resonates with the radio activists' notions about the community-level suitability of radio and its ability to foster transformative connections between neighbors.

This heritage played out in complex ways and had multiple implications for the transmitter workshop. Given their emphasis on personal and societal transformation, the activists did not wish to deny participants the experience of self-guided discovery, self-expression, or the formation of affective connection by too tightly controlling the workshop. On the other hand, if participants felt too impotent (or that what they produced was too inchoate), activists risked participants feeling as though their efforts had been wasted. Perhaps ironically, the transformative effects presumed to flow from technical engagement (imagined by both Appropriate Technology and participatory culture) were most elusive when neophytes were denied a structured experience in engaging with the technology.

Especially on the first day, the radio activists cultivated structure. As previously described, they offered a formal teaching track in which novices were taught a basic introduction to how a radio station works, with an emphasis on technical aspects of broadcasting. In addition, Brian and Jasper took pains to interrupt the engineers for formal reporting of their diagnostic activities. They insisted on a recap that included time for novices to ask questions (including very remedial ones). This move toward structure kept the novices occupied and included, providing them with tasks and roles to play. It also offered the novices opportunities to speak and participate without fear of being judged as ignorant or a hindrance to the diagnostic and repair mission that was the ostensible goal of the workshop.

At the same time, structure was inimical to other goals of the workshop. On the most basic level, the radio activists sought to provide participants with transformational experiences.[43] The imposition of structure could potentially render the activists unaccountable to participants' interests or values (especially to their exploration or creative expression).[44] At the same time, structurelessness ran a risk of producing disabling chaos or preventing the experience from having enough coherence to enable purposive engagement and expression. Novices were most frustrated at times when they were without prompts; "doing it themselves" when they were unequipped in terms of expertise was not ideal for them. And yet, the radio activists were loath to impose too much coercive control over the event.[45] This would have seemed to run against many of their organizational values and their strong collectivist ethos.[46]

Expertise was a significant issue in the interplay between structure and emergence. In an interview, Brian was critical of the culture of exclusion traditionally prevalent in engineering.[47] He summarized his occasional attempts to "manage" engineers working on technical projects with Prometheus: "while I don't explicitly say, 'Stop being a patronizing asshole,' I have tried to communicate that."[48] The problem went beyond merely keeping engineers from turning off novices by "being patronizing assholes." Indeed, the staff activists, not to mention the novices, needed the engineers if they were to make headway with arcane technical problems such as those they faced with the broken transmitters. Though activists valued self-organization and nonhierarchical participation, differentially distributed technical expertise threatened to exclude novices and erode the potential for "collaboration" that the activists embraced.

Conclusion

As anthropologist Jeffrey Juris writes, "activists increasingly express their utopian imaginaries directly through concrete organizational and technological practice."[49] The transmitter workshop exemplifies purposive engagement with technology that is oriented toward an emancipatory politics of broadened participation and agency. The radio activists' attention to technical practice and affective work around the meaning of artifacts exposes their mission as propagators of radio technology.

What this episode shows is that the "project" of technologically oriented media activism is complex. The work objects (or products) of radio activism require thoughtful interpretation, because the material practices surrounding hardware (the ostensible focus of this technologically oriented activism) do not fully capture what the activists seek to "produce." The transmitter workshop also confounds our understanding of whether "outcomes" are important. Participants might experience an intangible sense of gratification from the formation of community or political agency. For activists, a central tenet of their mission was to orient people towards a new politics of technology and expertise. Novices may not have come away from this workshop knowing a great deal more technically. But perhaps they may have learned to identify as people with the agency to fix a problem, to "participate." This was the activists' hope, anyway. But the fact remains that to really learn enough to build a radio station, novices would need much greater dedication (on their own, at tinkering meetings, and at barnraisings). On some level the activists recognized this. Even though they gave primacy to detailed technical know-how, they repeatedly came back to the idea of technical demystification as a form of political awakening. What was "really" being built at the workshop was a sense of coherence for activist projects (even those only imagined or far removed from the local setting). An affective relationship to activist technical practice was also being built. Both of these harmonize with the radio activists' goal of propagation of technology.

As they sought to provide workshop participants with transformative experiences, the radio activists stumbled over the issue of which dynamics could best serve their political goals. Some foreshadowing may be heard in the incommensurability between New Communalist and New Left approaches to politics. Like the New Left, New Communalists (who shaded into and overlapped with the Appropriate Technology movement in important ways) also sought to issue a challenge to the dominant social order. Yet they largely distrusted mainstream political activism as a route to this and instead sought "authenticity" and gratification through collectivized consciousness.[50] The prefigurative politics of Appropriate Technology and, later, "participatory culture" are mismatched to the more deliberate politics of groups such as media reformers. Prometheus was in a curious and contradictory position between these political modes in their effort to enact politics through technical enculturation. On the one hand, they believed that novices might have authentic and transformative experiences if simply given the freedom to conduct self-guided exploratory inquiries into the technology. On the other hand (and perhaps ironically) without structured routes to participation for novices, the group's activities took on unintended hierarchical dimensions. This ran counter to their goals for egalitarianism. In the New Communalist case, the communes' rejection of formal politics led to their falling back on norms from mainstream society.[51] The radio activists were politically committed to a different outcome. Prometheus recognized the extreme difficulty in overcoming these issues through a nominally nonhierarchical "self-organizing" mode of production. As a result, they adopted a formal consensus model in 2007 (after the period of this fieldwork). The activists found that they could not sustain their organization without forming a structure, but the "structure" they latched onto was a nonhierarchical one.

Collective collaboration did not solve the "problem" of hierarchical organization that is often assumed to be a feature of bureaucracies (but not of peer production networks).[52] Nor did it confer an automatic sense of gratification on all participants. The novices' experiences show that there are good reasons to be wary of romanticized notions of voluntarism and participation (especially in the realm of technology). Novices needed guidance and could not easily shed novice status. Peer governance in itself is not necessarily liberatory.[53] Before we celebrate the dynamics of self-organization, we should consider the difficult and elusive work of building and maintaining structures of participation, especially egalitarian participation.[54]

The transmitter workshop demonstrates that it took work to bind politics to artifacts. These transmitters, and the practice surrounding them, did not have an inherent meaning. For activists, engineers, and novices, the objects of focus and interpretation during the workshop activities varied subtly at times throughout the weekend. The workshop itself contained shades of ham radio's community building (but with a political cast that many hams would have found distasteful). Tinkering was meant to cultivate a sense of solidarity with media activism more widely. Yet the affective pleasure activists believed surrounded tinkering was hard for novices to attain without structures to guide them. Collective collaboration was insufficient to provide transformative experiences for novices or "awaken" them technically or politically. The experience and political significance of activist technical practice had to be actively constructed and maintained.

5 Fine-Tuning Boundaries

In summer 2005, attendees of the National Conference for Media Reform (NCMR) in St. Louis held an impromptu demonstration decrying the fact that *Democracy Now!*—a Pacifica news program featuring journalist Amy Goodman—was not carried by any St. Louis broadcasters. They walked away from the downtown conference site and set up a rally with banners and signs near a highway off-ramp. One Prometheus staff member spontaneously decided to get drivers' attention by writing "Democracy Now!" across her belly with a magic marker and flashing her bare breasts and stomach at the passing cars.

Whether or not this succeeded at capturing the attention of the passersby, her action seemed, on some level, a protest against the tone of the conference itself. More radicalized factions (including Indymedia activists) voiced criticism of the policy reform agenda that was a focus of the conference.[1] Prometheus organizers were caught betwixt and between in this setting. Well past their pirate origins, they were now marked by the legal status and institutionalization of LPFM. Nonetheless, Prometheus tended to identify with more radical positions. They often exhibited a critical stance toward "mainstream" nonprofit organizations, including Free Press, the sponsor of the NCMR. Indeed, the bare-chested Prometheus activist seemed to translate some of her frustration at the conference into an oppositional and "improper" bodily response. (*Democracy Now!* staff, including Goodman, looked on while this took place, and I do not know whether or how they reacted internally to this addition to the rally.) Needless to say, this mode of expression would certainly be off-limits if a Congressional meeting about LPFM were to go badly. But maintaining decorum at NCMR was not a priority. In fact, this breach in composure served to mark a boundary between Prometheus and the "media reformers" (exemplified by Free Press and Beltway allies in media democracy).

In this chapter, I explore Prometheus's work during this pivotal moment, characterized by the institutionalization of LPFM and the maturation of Prometheus as an organization. How did the radio activists grapple with becoming institutionalized? What resources did they draw on to manage anxiety and maintain a coherent identity in the face of varied work and a changing organization? And in what ways did their technical activities get brought to bear? Although the radio activists' work was, in actuality, often mundane, they made consistent oppositional displays. They downplayed their move toward becoming a sustainable nonprofit and played up their alterity vis à vis mainstream nonprofits and Washington, DC–based policy insiders. In addition, they invoked their technical work in a way that drew a contrast between themselves and "mainstream" reformers and nonprofits. In earlier chapters, I attempted to show how technical affinity was seen as a way to liberate novices. But here the radio activists embraced it to symbolically unshackle themselves from the drudgery of filing and database maintenance or—even more troubling—policy advocacy. This chapter shifts focal range, zooming out from barnraisings and workshops to trace the radio activists' experiences and relationships as they extended out into the world. In Prometheus's interactions with "outsider" groups, we can observe what sociologists call "boundary work," in which the group sought to differentiate itself from others in the terrain of media democracy work.[2] At the same time, the radio activists also used identity work to efface potential differences within their organization.

Activist Work: Inside and Out

Prometheus's office was a low-tech work environment. It didn't look like an assembly line or a science lab. Despite the ostensible familiarity of the office setting and tasks, it was not self-evident to the observer what the point or product of the work performed there was.[3] And yet the activists experienced their work tasks as meaningful; in particular, they did not experience the work environment's variability as incoherence. The following descriptive sections are meant to illustrate and describe workplace activities and provide vignettes that demonstrate some "typical" work situations and undertakings at Prometheus. However, the categories I use to divide these activities (paperwork, organizing work, and technical work) are not cleanly delineated from one another. Indeed, breaking up the work into these categories presents some problems. In many cases, work activities I have placed in one section could easily fit into another. Was a barnraising an organizing activity first and foremost or was it technical, in that technical work occurred and it was seen as a service to license-holders? Were the handbooks that Prometheus provided to stations paperwork, or were they technical work because they offered specific suggestions to station-holders based on "data" that had been compiled? Were the handouts Prometheus produced to distribute at talks, legislative visits, and demonstrations paperwork or organizing work? Does this matter? My intent in proffering these descriptive sections is to provide empirical, ethnographic material that grounds radio activism in the close study of work practices and products. Yet none of these *descriptions* of work practices meaningfully captures what Prometheus's work *is*.

"Work is communicative action, which can therefore be analyzed in terms of performance," writes Andreas Glaeser.[4] Activists' undertakings can be understood as actions in the world that communicate meaning to other members of their work world and to parties with whom they interact. Occupations themselves are "things of boundaries."[5] In other words, members of professions demarcate territory and define what is and what is not within the bounds of their profession.[6] The notion of jurisdictional boundaries is useful in the analysis of identity work within occupations. Some workers are necessarily "inside" the boundary of occupation, whereas others are excluded from this designation.

In this chapter, I deliberately alternate among the terms *worker, organizer,* and *activist* to refer to the people who were paid to work at Prometheus. I call unpaid Prometheus workers *volunteers.* Elsewhere in the book, I am more relaxed about these distinctions and commonly lump these people together as "activists." But in this chapter I want to highlight that it should not be taken for granted that "activism" is a unified category of activity. Occupational titles tend to be stable and to imply specific ideas about the characteristics of workers. But the radio activists were themselves quite free with their titles, sometimes using *organizer* and other times *director* or *coordinator.*[7] The paid staff was arranged to be nonhierarchical, and they eschewed titles such as *executive director* (though at one point Jasper joked that the other staff would like it if he adopted that title because then they could organize against him).[8] The fluidity of the titles that Prometheus workers assumed spoke to a relative discomfort with being narrowly identified by tasks. It also reflected dissent from the mainstream occupational practice of having a stable occupational title. Prometheus staff tended to leave off job titles in their e-mail signatures. But they did use the organization's title in these e-mail signatures, for example signing off as "Ellen [Lastname], Prometheus Radio Project." When needed, they also adopted occupational titles: in a grant application, staff referred to themselves as "[Ellen], Program Director, [Thomas], Development Director, and [Brian], Technical Organizer."[9] More commonly, they deployed such titles flippantly and fluidly. A woman who had volunteered for Prometheus for years stepped up her involvement and asked about the title she should use. Jasper replied, "everyone at Prometheus gets to make up their own title. it is an initiation. how about Spectral Director? or Chief of Rainbows? Wave Associate? You can't have Director of Electromagnetism, that's mine!"[10] This quote draws out a number of salient themes of this chapter: the prominence of "the technical" for Prometheus; the activists' embrace of a wide diversity of tasks and refusal to reduce the essence of their individual and collective work to any single practice; and significantly, their willingness to claim seemingly unrelated or even contradictory tasks as all falling under the rubric of activism.

Kinds of Work

In chapter 1, I provided a general overview of Prometheus's work, including their priorities and the climate in which they were working as an organization prior to my time with the group. This chapter discusses the work I observed firsthand during my time with Prometheus, in the office as well as extending outward into other environments. It addresses what these varied forms of work meant. I divide the sections into "paperwork," "advocacy and organizing work," and "technical work." These categories of work are descriptive, rather than analytical. They are not exhaustive, and are not cleanly delineated in all cases. I return to an analysis of these work activities in the latter portion of the chapter.

Paperwork

A ubiquitous feature of many work environments is paperwork.[11] Prometheus workers were compulsive and manic writers, producing a variety of printed and electronic inscriptions.[12] The internal office email list was called "basement" (a reference to their office being housed in a church basement). It generated much discussion and many drafts for review. The office workers struggled to maintain growing electronic and paper files. Papers were constantly shuffling off desks, into the mail, and into the files. Workers wrote grant applications, fact sheets, spreadsheets, newsletters, budgets, meeting agendas, meeting notes, paper and electronic correspondence, and web pages.[13] Some documents were for internal use and review only; others "emanated" from the office.[14]

The production and management of paperwork (much of it electronic "virtual paperwork") cannot be overstated. One of my first interactions with the Prometheus workers, described in the book's introduction, was in 2003 when they were nervously awaiting the results of the MITRE Corporation's study on signal interference. They were concerned that if the results of the study in any way confirmed the broadcast lobby's predictions of interference, they would face an uphill battle to protect the existence of LPFM (let alone expand it). To prepare for this, they were marshalling their resources. They had prepared a request for proposal (RFP) to find yet another independent engineer to comb through the MITRE results. This engineer was to write an analysis of the findings "in a manner suitable for submission as comment to the FCC in a timely way to meet all deadlines for public comment in the proceeding."[15] When I was meeting with Prometheus for preliminary discussions about my project, Jasper strongly suggested that I could work on the RFP project, too. He thought it would be of interest to me as a question of sociology of science, and he was keen for me to be on call to help with writing needs. I came away from this meeting not unwilling to help. But I also felt as though I had experienced a classic anthropological encounter in which I was enrolled by the "natives" in service of their own priorities. I was somewhat less certain how mine as a researcher were being met.

Soon after these conversations, the MITRE study was released. It upheld the engineering results put forth by LPFM proponents, and thus obviated the need for LPFM advocates to challenge it or commission new research. Almost immediately, Prometheus produced a press release stating "Broadcast Lobby Caught Red Handed with Red Herring—'Oceans of Radio Interference' Proven to Be Puddles by Independent Study of LPFM." Rather than having to assume a tone that was technical or appropriately "neutral" for a policy debate, the release included this rather sassy quote:

> "I hope that the wild goose chase for interference—and the claim that a dinky hundred watt community station can cause this kind of problem for a 20,000 watt commercial station—can finally come to a close[,]" said [the] Technical Director of the Prometheus Radio Project. "I know some lobbyists at the National Association of Broadcasters may not know what to do without Low Power FM radio to beat up on anymore, but I'm sure they can find gainful employment searching for other imaginary things like African uranium shipments to Iraq."[16]

Anthropologist Kim Fortun discusses the skills she acquired during her fieldwork on activism after the Union Carbide disaster in Bhopal. She writes that "there is a constant need for words; satisfying one demand provides only short reprieve before the call to language comes again.... New things had to be said, but through a negotiation with older forms."[17] Her description is apt; as the MITRE example demonstrates, Prometheus workers were constantly poised to produce language and documents that quickly responded to the changing terrain in which they operated. They were always at the ready to change tacks and rhetoric as soon as they perceived a shift in the discourse available to them.

Prometheus organizers routinely produced newsletters, flyers, and handouts. The newsletter was an important document. Its ostensible purpose was publicity, and especially to appeal for donations to people already sympathetic to Prometheus's mission. But it also served as a site of storytelling; it was a display not only to readers but to the workers themselves. The newsletter typically included an update on Prometheus's legislative activities, as well as notices of upcoming barnraisings and other events. FCC-related news was also of potential interest to the readership. Newsletters also included a "Tech Corner" column and contributions from guest writers.

The Prometheus newsletter was text dense. But images were selected for their potential effect on the audience or that were particularly representative of Prometheus's work. For instance, this newsletter's cover image was remarkable for the way it constituted a performance or storytelling of Prometheus's values to itself and readers. Its use of a homespun drawing, of two punk-rock women listening to an old-fashioned cathedral radio, was intended as a striking portrayal of the vitality and alterity of this old medium (see figure 5.1). For documents and the newsletter, emphasis was placed on finding good images and not having them overrun by text.[18] A typical handout was a bright neon-yellow 8½ × 11" photocopy that read, "Action Alert! Grab Your Bolt Cutters—It's Time to Unshackle Low Power FM!" Though its text pertained to Prometheus's policy work, visually, the document centered around an image of a boltcutter.

The purpose here is not to catalog and discuss every form of paperwork in the Prometheus office, but instead to draw attention to the acts of producing and managing paperwork in the office setting. A final category of paperwork that warrants attention is of a more administrative nature. During my fieldwork, Prometheus was routinely concerned with information management and retrieval. The organizers wanted to create a more logical, sustainable set of rules and systems for dealing with papers in service of greater transparency, sustainability, and accountability as an organization. During my time with the group, the workers researched health and liability insurance, learned and implemented a new accounting system, attained 501(c)(3) status, and endlessly worked on the issues of website and filing redesign.

Following are two examples of attention to these more "administrative" functions:

Figure 5.1
"Prometheus DeLivered" newsletter (Winter 2004–2005).

One e-mail containing agenda items for a meeting stated:

1. Do we want health insurance? Do we want general liability ($500/year) and D&O ($750/yr) insurance? 2. Going over said policies & options, in particular health insurance. 3. Deciding who we'd like to broker our health insurance for us. Then we can go ahead with it.... Once we sign up for a plan, we have 10 days to review the policy.[19]

And another e-mail to the staff from someone who volunteered (unpaid) and consulted (paid) to assist Prometheus with special projects:

Hi Prometheans! About 6 months ago I set up a nifty filing system for all the amazing literature and other distributable materials that you keep flowing among the media activist community. Please take a quick moment to help me assess how it is working and make any necessary tweaks....

1. have you used the materials files to retrieve a document in the past few months?
2. if so, did you use the [E]xcel [program] "map" to find the file you needed?
3. have you added a file to the materials files in the past few months?
4. if so, did you add this to the map or just to the hard files?
5. did you try to do one of these but not have luck? if so, what was confusing or difficult?[20]

These mundane examples were typical in that they represented internal office communications about administrative functions; I am not analyzing them at any greater level of detail here. In spite of activists' valorization of barnraisings as their "main" work, they expended a great deal of organizational energy on issues such as health insurance and filing. I do not argue that these e-mails are wholly representative of communication within the office, but there was a heightened attentiveness to more "bureaucratic" concerns during this period.

This was evident at a staff meeting I attended. During a discussion about hiring summer interns, the staff debated about how many people to bring in for the summer and who could work on which upcoming projects. Jasper was fairly committed to the idea of bringing in a Canadian woman with community radio experience who wanted to help plan an upcoming barnraising event. Thomas, the staff member who worked on development, accounting, and other administrative issues, was nervous about hiring her because of her questionable status as a legal worker. Jasper brushed off Thomas's concern about this, but Thomas would not drop the matter, saying, "It's not a question of *if,* we're *gonna* be audited this year [due to our 501(c)(3) application process], and I don't think I can bury a check or even cash." The two compromised on hiring her under the assumption that they could probably "bury" a food and housing allowance, but not wages. It was very clear from this exchange that having to consider things such as the paper trail surrounding wages, taxes, and the legal status of a part-time worker was not routine for them. The significance of this increased attention to bureaucratic and administrative details in their work is discussed later in this chapter.

Advocacy and Organizing Work

The next category of work performed by Prometheus has to do with their advocacy (primarily in Washington, DC) and organizing work (in which they focused on members of the public, the "grassroots" base).

I. Washington, DC

In February 2005, Prometheus members held a meeting in the offices of a liberal advocacy nonprofit, Common Cause. They were there to teach members of the public to lobby their elected officials (hoping to build support in Congress for the bill to expand LPFM, as discussed below). Prometheus itself could not legally participate in formal lobbying activities, but they had allies in Washington who attended the meeting to give Prometheus and citizens some lobbying tips. In the meeting, both Prometheus and the lobbyists reinforced the notion that Prometheus was in a role between that of "ordinary" citizen and lobbyist. Ellen told attendees that "we hate that we have to get you up to do this and undo the bad work of the National Association of Broadcasters—we'd rather concentrate on expanding LPFM and building more stations, but we have to do this first."[21] A lobbyist emphasized the role of the citizens: "Without [you] people here in the field, [my organization is] just three smart lawyers talking to ourselves."[22] Both lobbyists and Prometheus members were on hand to accompany people on legislative visits; people were told not to worry because they would have an "expert" with them, but Prometheus did not exactly see themselves as possessing expertise in this domain. Ellen described Washington, DC, as a "crazy place full of magical buildings and towers that we don't understand all that well."[23] And Jasper remarked, "The problem with Washington is, Washington is weird. I don't know how you do anything there. It's all secret and you have to be very powerful to get anyone to listen to you."[24] Thus Jasper and Ellen emphasized their ostensible lack of familiarity with DC, belying the expertise they in fact were coming to possess over time.

This lobbying exercise was held in conjunction with an event called "LPFM Day," hosted by the FCC. This event constituted another example of Prometheus's work as policy advocates. The FCC hosted a half-day "Forum on Low Power FM," attended by Prometheus and LPFM station holders from around the country. This was largely a symbolic gesture on both sides. Advocates knew the FCC was not going to announce any changes to rules that would positively affect LPFM. This was partly due to the fact that then-chairman Michael Powell was on his way out of the commission, and as a result progress on various proposed rules had stalled. At the same time, it was an act of good faith for activists and broadcasters to attend and remind the FCC that they still sought improvements to the LPFM service. There was also a sentiment among advocates that they did not want to be ignored by the FCC. All five commissioners including Jonathan Adelstein, Michael Copps, Kevin Martin, Kathleen Abernathy, and Chairman Powell were in attendance for at least part of the forum.

The panelists told the commissioners about their experiences with their stations and made claims about the value of LPFM. A station holder from Oroville, California, told the commissioners that "LPFM stations do something that big stations can't or won't do—they open up the community and allow us to talk to one another."[25] Another panelist called on the FCC to "overprotect" LPFMs from encroachment and interference, telling the commissioners to "take on the NAB and do your job." He likened the situation of LPFM to the story of David and Goliath. Panelists made requests for large and small changes to policy that would ease the maintenance and governance of LPFMs. They sought protected status for LPFM against encroachment by full-power stations. They also spoke out against giving primacy to translators (repeater transmitters that relay signals of content originating elsewhere). As things stood at the time, LPFMs had secondary status and would thus be first to lose the right to broadcast in a spectrum dispute with a translator, even though LPFMs were likely broadcasting original content. The FCC remained coy over whether the larger considerations requested by LPFM advocates would be found to have merit. Nonetheless, the opportunity to enter the halls of power and command the attention of the commissioners for half a day was an important exercise for the Prometheus organizers. They had put significant effort into selecting the people who appeared on the panels. They had also drummed up supporters who lived near Washington to attend the event, ensuring that no chairs in the room were empty (I would estimate that around two hundred people attended this event, though I do not have an official count).

At the same time, Prometheus worked on advocacy for bills to protect and expand LPFM. In the Senate, the Local Community Radio Act of 2005 was being introduced by senators John McCain (R-AZ), Patrick Leahy (D-VT), and Maria Cantwell (D-WA). Representative Louise Slaughter (D-NY) introduced similar legislation in the House. These bills called to negate the Radio Broadcasting Preservation Act of 2000 (passed in an appropriations rider). (LPFM was restored when President Obama signed the Local Community Radio Act of 2010.) The need to present LPFM as bipartisan was strong in the Republican-dominated Congress of 2005. Advocates were fairly successful at portraying LPFM as a nonpartisan, big-business-versus-local-community issue, highlighting its appeal to "churches and schools."[26] LPFM advocates considered Congress members of either party with strong ties to the broadcast lobby to be enemies, regardless of party. But they recognized the need to court right-wing politicians in particular, hoping they could make appeals on the basis of localism or the perceived relationship between corporate media consolidation and indecency in broadcasting; opposition to consolidation could be found on the left and on the right, though not always for the same reasons. Particularly in rural areas (where many new LPFM licenses had been granted), evangelical churches were an organized force that desired LPFM stations and could lobby their representatives. In a 2005 article in *The Nation,* Rick Karr wrote, "Prometheus is trying to organize the fundamentalists. 'In order to be able to expand low-power FM on the Hill, we're going to need these people,' [said an organizer]."[27]

Yet the US Congress was so divided in the 2005 sessions that the issue of partisanship remained overwhelming. Washington insiders close to Prometheus were actually dismayed that Representative Slaughter was introducing her legislation because they feared that without an effort to secure bipartisan sponsorship before introducing the bill, it was a wasted effort. The rationale given by these lobbyists was that in such a divisive political atmosphere, there was no way that Slaughter, a Democrat who was not on the Telecommunications Committee, would be able to gain enough allies to pass the bill. Lobbyists lamented that Congress was "like junior high school" in its politicking, but underscored that this was how Washington business was done. It was clear that the merits of the issue aside, Slaughter's bill would be unable to gain sufficient support. One lobbyist stated, "You can talk to as many people as you want to but it all really rests on one or two [representatives]."[28]

With this as a backdrop, I accompanied Prometheus workers on a trip to Washington in which they met with the House Progressive Caucus of the Democratic Party about the LPFM legislation circulating in the House. The caucus meeting was held in a private dining room in the Capitol building. The group contained about five representatives and the staffers of another fifteen representatives. One of Prometheus's lobbyists, John, opened the meeting with a description of the history of LPFM. I noted that he began in 1999, and left out the history of unlicensed micro-broadcasting and the legal skirmishes between the FCC and "pirates." Addressing the caucus, John put the issue in simple terms, framing it as one of a public being stymied by red tape and foot-dragging regulators: "community groups ... just want stations, [and they] accuse the FCC of hoarding spectrum. [The community groups] say [LPFM] is technically possible, if only the FCC would update technologically."[29] Jasper then took over, stating that LPFM "[provides] an important opportunity to take representatives and senators to task for siding with corporations and against constituents like small local conservative churches."[30] He advocated for media in the hands of local groups and condemned the continuing expansion of corporate media. He also intimated that the Progressive Caucus should take a special interest in promoting LPFM, because Prometheus and the caucus shared the goal of getting progressive media outlets into the hands of as many community groups as possible, in cities and towns across the nation.

In this context, Prometheus represented themselves as explicitly in favor of building up a left-wing media apparatus. Ellen stated this outright, saying that Prometheus's goal was to get the bills passed while organizing to have progressives first in line for new stations.[31] In other contexts, however, they were far more guarded about this goal. Indeed, there was a tension here between their rhetoric about citizens across the political spectrum "having a voice" and their radical left-wing ideals (the extent of which they perhaps would not even fully disclose to the Progressive Caucus members).[32] The issue of what radio *does* was unresolved in much of the radio activists' ideation and practice: was the power of radio inherent in the artifact? or did it matter who was speaking, what was being transmitted? (Chapter 6 addresses this issue most directly.)

On visits with staffers, Prometheus's tone ranged. With some staffers who were known to Prometheus and who were considered sympathetic and savvy, (such as a staffer for Representative Jan Schakowsky [D-IL]), the conversation was strategic and specific. They focused on winning technical changes (alluded to in the discussion of the "LPFM Day forum") at the FCC and working to gain support for broader expansion of LPFM in Congress. Schakowsky's staffer agreed with John the lobbyist, stating that "with [Representative Slaughter's] name first [on the bill], it's not moving—jurisdiction is a big issue here."[33] I noted that this staffer seemed very sympathetic to LPFM. Schakowsky's district (part of Chicago) would be unlikely to receive new LPFM stations even with the passage of the desired legislation because of spectrum crowding. But the staffer mentioned "pirates" operating in Chicago, and it was clear that she supported them (or was at least happy to look the other way). By contrast, the issue of unlicensed broadcasting was omitted from the discussion during Prometheus's meeting with the Progressive Caucus.

In other meetings, the staffers knew virtually nothing about LPFM. Prometheus workers politely described the issue as one that no reasonable person could oppose, stressing its bipartisan appeal and the grassroots demand for locally controlled small-scale radio stations. In return, the staffers treated the activists very civilly and evenly, but studiously avoided signaling support. They instead speculated about other congressional offices where the activists might get a sympathetic hearing. Repeatedly, staffers politely raised the issue of the bill not having gone through the appropriate committee.

A final meeting occurred with Representative John Dingell's (D-MI) staffers, in which the Prometheus workers were more cautious, yet also very specific and strategic. Dingell had in 2000 supported the Radio Broadcasting Preservation Act that sought to limit LPFM. Prometheus hoped to find out from his staffers whether he might be willing to reverse his position and endorse the new legislation (or at least obtain his word that he would not oppose them again). Lobbyist John, Prometheus workers, and former FCC commissioner Gloria Tristani attended the meeting. Dingell's staffers were careful not to guarantee anything but did offer vaguely encouraging words. The Prometheus organizers were polite and somewhat subdued. John did much of the talking. He said that the issue could be "sold" to Dingell without seeming to contradict his earlier position, if it were framed as "the FCC did their job [in commissioning the MITRE study and getting independent engineering approval for its recommendations for LPFM] and now it's time for Congress to do theirs." He also told the staffers that he had been "tracking" the issue and getting "no significant pushback." By this, he meant that he was convinced that the broadcast lobby was not gearing up for another big controversy around the issue. He thought this fact might also make supporting the LPFM bill more appealing to Dingell, as it might allow him to quietly support it without risking a flap over the reversal of his earlier position. A staffer asked whether the LPFM advocates had met with the NAB. John said he had, and that his sense was that the "rank-and-file doesn't really care."[34]

I sketch these private meetings with staffers, as well as the one with the Progressive Caucus, in order to illustrate the experience of Prometheus workers in Washington as advocates. The lobbyist, John, stepped up to speak in situations that seemed more "sensitive" or to that required the specialized perspective and language of someone "inside" Washington (e.g., "pushback"). The Prometheus organizers played a slightly different role when they spoke very frankly with the staffer who was perceived to understand and support LPFM. At other times, they showed deference to the Washington process, stressing that their own expertise lay elsewhere, especially in their ability to organize and represent the "grassroots." These instances seemed to demonstrate that Prometheus did not draw heavily on Washington "insider" expertise (evident in meetings with staffers and in the Common Cause workshops). Instead they played up their own sense of being different from people who were more able to deploy this "insider" expertise. In this context, it is worth noting that the interventions of "laypeople" in expert decision-making processes may make it difficult to straightforwardly determine "who is a 'layperson' and who is an 'expert.'"[35] The categories of who is "expert" and which expertise(s) may be said to "count" are not clearly defined. These categories may also shift over time or across contexts.

II. On the Road

It would be very misleading to represent a majority of Prometheus's "organizing" work as having occurred in Washington, DC. Since the earliest days of the group, members relentlessly toured the United States to speak to community groups, independent media activists, school groups, and the like. I spent nearly a month on the road touring with Jasper before a barn-raising event in Tennessee. We began in Chicago and traveled to Round Lake, Illinois; Ligonier, Indiana; Bloomington, Indiana; Richmond, Indiana; Urbana, Illinois; St. Louis, Missouri; Lexington, Kentucky; Louisville, Kentucky; Nashville, Tennessee; New Tazewell, Tennessee; Knoxville, Tennessee; Asheville, North Carolina; and Carrboro, North Carolina; before heading back to Philadelphia. We spent one to three days in most of these places. The rationale for this tour was to visit LPFMs in the region with whom Prometheus had a relationship (even if it was just talking on the phone a few times). Jasper spoke at stations, community centers, and colleges along the way in an effort to raise the visibility of Prometheus's work on LPFM. He also encouraged people at the events to attend the upcoming barnraising, which was to be held a few weeks later just outside of Nashville, Tennessee.

A typical presentation, given to college students on the campus of Earlham College (a Quaker liberal arts school in Richmond, Indiana) included the usual narrative history of the Prometheus group and its activities. It included Jasper's realization that media access and media coverage were core issues in the range of social justice causes in which he had worked before community radio. He also highlighted the Radio Mutiny pirate collective, changes in telecommunication policy (the 1996 act permitting unlimited ownership of radio stations across the nation, and the introduction of LPFM in 2000), Prometheus's lawsuit against the FCC, and their barnraising events, including international work in Guatemala and Nepal. Jasper gave his presentation in an informal and humorous way, poking fun at big business and the government. At another talk, Jasper addressed an audience largely composed of former pirates from KFAR (Knoxville First Amendment Radio). The station had been on the air for three years, but had been shut down by the FCC just a few months before our stop in Knoxville. Jasper was even more irreverent in this setting. Describing Prometheus's strategic work to move the legislation through Congress, he said, "If it's sponsored by a Democrat, it might as well be a lesbian anarchist from Mars!"[36] At each event, we sold t-shirts and books and distributed literature. We also used a series of 8½ × 11″ mounted photos to illustrate Prometheus's international station-building work and provide a visual component that bore witness to his descriptions (see figure 5.2).

The other main activity on the tour was visiting stations. After the talk to the KFAR audience in a Knoxville restaurant, attendees went back to

Figure 5.2
Two women soldering a transmitter board in Nepal. Slide from Jasper's trip to attend a conference sponsored by AMARC, the World Federation of Community Radio Broadcasters. Volunteer photo.

the KFAR studio, which was located in a trailer on a ridge above town. We chatted with the people who had been at Jasper's presentation and with the people in the studio. (The studio was still in use for webcasting and had been rechristened CROK, or Community Radio of Knoxville.)[37] Some of the local people knew Jasper already from his earlier tours or activist media gatherings. It was clear that they were honored to have him visit. I was all but shocked by one woman's display of devotion to Prometheus. Jasper and I arrived at the KFAR/CROK studio well after dark. Yet (even in the evening dimness) she happened to notice that the station's Jolly Roger flag (hanging from the antenna tower) was wrapped around a tree branch and not hanging freely. To Jasper's and my amazement (and no little horror), she climbed the forty-foot tower in the dark to unfurl the flag, saying that she was embarrassed it was not on display properly for Prometheus's visit. Fortunately she made it back down in one piece; back on the ground, she remarked to herself that her climbing coach would be unhappy to learn that she had climbed the tower without her harness.

We slept that night at the home of KFAR/CROK volunteer Ian, whose day job was running a home security and automation company. While driving to his house, we tuned the car stereo to a radio signal emanating from his basement. There is some interest among community radio aficionados (and "techies" of various stripes) in what are called "part 15" transmitters (see chapter 7). These are legal, unlicensed, very low-power transmitters that typically have a range of a few hundred feet. Ian had set up a part 15 transmitter that was being fed audio from a web-stream. He was broadcasting international news from his home, which could be heard from a couple of blocks away. He was proud of having tinkered with the transmitter to boost its signal to somewhere between 1,000 and 1,500 feet. Car stereos typically have good receivers, so that helped us receive the signal, too. At his home, he showed us his apparatus. He was excited about the possibility of setting up a "cloud" of part 15 transmitters to relay in order to repeat a signal and blanket a larger area. Jasper listened to him very patiently and asked some technical questions, offering some tips. But we were tired and happy to head off to sleep when we could. Ian seemed happy to have company and eager to entertain. He offered us homemade beer (which we accepted) and marijuana (which we declined).

Other LPFMs we visited were much less piratical and more "middle-America." Ligonier, Indiana was the site of the Indiana Historical Radio Museum and WNRL-LPFM (unofficial slogan: "We're Not Real Loud—but We're Real Good"). The museum was located in an old filling station. It housed vintage and novelty radios spanning the history of radio, as well as an assortment of original audion tubes (the amplifying vacuum tube developed by Lee DeForest in 1906). We met with a father-and-daughter team who were responsible for the creation of the station and museum. The father, Jack, owned an electronics shop and worked as a state trooper before retirement. He had collected radios on the side for decades. Although Jasper and Janine had corresponded many times over the years spanning WNRL's application process, this was their first face-to-face meeting. Both very happy to finally meet each other in person. Janine had worked on a small, technical problem with her LPFM application that the 2000 change in LPFM regulations had caused. She had spent a lot of time on the phone with Jasper and the FCC during the process of obtaining WNRL's license. We spent most of the visit with Jack, because Janine was quite ill and did not have much stamina for visitors. The museum and radio station both subsisted entirely on individual, corporate, and municipal donations from the local Ligonier community. The station's transmitter was located at the public high school, with the studio in a municipal recreation center.

Underscoring how intimately the people volunteering at the station knew each other, the programming schedule listed only programmers' first names and nothing about their shows ("Fritz," "Beatrix," "Jose N.," "Reserved for ball game," and "School use"). I was told that these programs included local news, sports, music, and Spanish-language talk.

These station visits were not "merely" social, of course. In every meeting with LPFMs, Jasper was careful to ask questions about how the stations were faring and what challenges they were facing. Challenges included encroachment and interference by full-power stations, difficulties with fund-raising or station governance, and meeting FCC requirements pertaining to technical specifications and board constitution. In turn, Jasper would share news of Prometheus's Washington efforts to pass legislation and effect changes in the FCC's regulations that would protect LPFMs. He also encouraged station-holders and volunteers to come to the upcoming barnraising in Nashville in order to meet and "network" with other members of LPFMs in the region.

Drumming up barnraising attendees was crucial for Prometheus. Activists tried to provide new LPFMs with strong ties to existing LPFMs and other community stations in the hopes that these groups would share knowledge and support one another. Prometheus wanted LPFMs to think of themselves not only as part of a network of stations whose interests and concerns were interconnected but also as members of a media democracy movement. The Prometheus organizers routinely stated that the barnraisings were less a means to build radio stations and more a means to promote solidarity and action; they hoped to foster and sustain a movement. We can hear evidence of their role as propagators here: the organizers sought to diffuse the technical artifact of LPFM and to define its meaning, including its role in promoting democratic social relations.

The other major activity that deserves attention as an organizing routine is the barnraising event. In contrast to much of the activists' work, barnraisings were extremely visible and symbolically important. They represented a space for demystification of technology and political empowerment, providing a place for the "barnraising "ideal to come alive—people joining together to tackle a project that an individual or small group alone would struggle to accomplish, with an emphasis on interdependence and cooperation. The barnraising was a major site for Prometheus to perform and make visible their mission. (Because barnraisings featured prominently in chapters 2 and 3, I do not additionally describe them here; I do, however, wish to emphasize that in spite of the range of activities occurring at barnraisings, their value as an *organizing* activity was paramount for Prometheus.)

Technical Work and Other Services to LPFMs

Prometheus also conducted various forms of technical work. They provided services to groups who needed assistance and dispensing advice on a wide range of issues. (Here I am defining "technical work" not only as "related to electronics hardware," but also as providing guidance that is "arcane, detailed, and of interest to a limited discourse community of LPFM station holders or prospective station holders." Much of this "technical work" also happens to be "technical" in the sense of "related to electronics, machines, or technology.") Although a plan for station governance and a handbook on managing station volunteers are not "technical" in the sense of relating to electronics, these items are specifically related to running radio stations.[38] "Technology" is largely an actors' category: the activists used "technical" to refer to audio, computer, and radio transmission hardware and software related to the production of LPFM and community media.[39] As will be explored later in the chapter, certain "technical" practices were important for the group to articulate a position around "technology" and its meaning in the group, while other activities that were designated as "nontechnical" were less freighted with symbolism.

As previously indicated in the descriptions of the barnraisings, the transmitter workshop, and Geek Group, a good deal of radio activist work was hands-on, including tinkering with hardware. At barnraisings, technical expertise was furnished by Prometheus staff, Prometheus volunteers from all over the country, and local volunteers who were affiliated with the new radio station or who just wanted to help out. Prometheus did not charge groups money for barnraisings, instead carefully selecting the groups with whom to hold barnraisings and funding the barnraisings in collaboration with the local groups. But Prometheus also received requests for assistance from groups who were not barnraising partners and were unable to pool local low-cost or volunteer resources. Prometheus offered these organizations a fee-for-service program, in which they sent a few staff and volunteers to work with the local group, who in turn paid them for their time and materials. Prometheus still emphasized the training aspect of their mission and tried to structure the work days to include workshops. They encouraged the local group to have volunteers attend the work days so that they could learn new skills. Discussing a fee-for-service project, Jasper wrote, "[Our] proposal should emphasize the training aspect. There is no sense in us going and just plugging a bunch of things together. We can say that we would expect many people to come. I think we could get at least 3 or 4 people from the area who could conduct workshops in audio, station management, news and other things, so there could be one workshop running at any given time along with the studio construction."[40] This quote came from an internal e-mail in which the Prometheus workers were drafting a proposal for a fee-for-service project building an Internet radio station in Chicago (see chapter 7). Brian agreed, replying "since the station construction is relatively simple, the big reason for bringing Prometheus in is the participatory nature, and making this a 'mini barnraising' could be really great."[41] (Note the resemblance here to the transmitter workshop: the activists did not wish to squander pedagogical opportunities.) Prometheus wished to distinguish the services they provided from the standard practice of hiring an engineer or consultant to come in and build an entire radio station. (An LPFM might otherwise build or maintain their station this way.)

The rates Prometheus charged were fairly flexible. They would quote a "real" rate and encourage the service recipient to fund-raise for their fee. But they also planned for the need to substantially discount their rate if the recipient could not raise the funds and desperately needed assistance. In the same e-mail, Jasper wrote, "I think that the discount should be referenced to the ultimate source of funding. If they are doing it all out of bake sales, it is one thing. But if they get significant grant funding, we should charge a real rate. Let's put the proposal out at full price, and note that we can discount it significantly in accordance with ability to fundraise."[42] The fee-for-service projects would ideally cover their costs, but Prometheus considered services to community groups to be an important part of their mission. If they had to subsidize the cost of doing a project for a group in need, they would.

Prometheus also administered two e-mail lists that served as information clearinghouses and discussion sites for information about LPFM. One list, "Stubblefield" (named for Nathan B. Stubblefield, who experimented with wireless telephony before Marconi's famous displays of wireless telegraphy), was a general list. Topics ranged from LPFM legislation updates to discussions about fund-raising, details about compliance with FCC regulations, and FCC news. One station posted a request for sample budgets to aid them in writing grant requests. They received this reply from the list:

for insurance purposes we recently had to compile this equipment list:

EAS [Emergency Alert System] 911 T2 Encoder/Decoder $2820
EAS Interrupt Relay Unit $53.52
EAS 943 Telephone Access Unit $480
MT300-U 300 watt FM Transmitter $1950
Coax Connectors and Two bay antenna $946

ASI4215 Digital Audio Adaptor $241.25
Tascam 202MKII Cassette Deck $399
Aphex Comprellor $1200
2 Rolls FM tuners $298 each
Gemini Dual CD Playter $350
Behringer Headphone DA $150
2 Behringer Voice Processor $150 each
Generic Computer $800
Canopus ADC100 DV Convertor $250
~$10,230

absent are borrowed mics and a borrowed field mixer, so no console in this price list, plus we have another computer in use now. you can get a cheaper EAS unit but the city paid for part of ours. we didn't skimp but we didn't exactly go for luxury either, and of course now we want more, more, more—good mics / booms / turntable mixer / etc.[43]

This mundane e-mail highlights a couple of important features of a typical LPFM: a fairly minimal and bare-bones studio setup (e.g., "generic computer") that could be improved if funds became available, and partnership with the municipality in order to cover the cost of the required emergency alert system (see chapter 6 for more on the issue of LPFM and public safety). This post represents well the type of specialized content routinely found on this list's traffic. It also gives list members (and the reader) a sense of the capital investments required for an LPFM station.

Another typical e-mail from this list, sent by a Prometheus board member, warned LPFM stations to be careful about the issue of underwriting. LPFMs by law must be noncommercial and not for profit. They may not broadcast advertising:

There are very clear prohibitions against sponsored programming on noncomm[ercial] radio.

you are allowed to air paid announcements for other Not for profit groups, but if you are taking underwriting for a program, the kind of language you are legally allowed to use to acknowledge their gift has restrictions. this is not to mention that there is an inherent conflict of interest.
you can approach these businesses for general support, and say that "this part of the broadcast day has been brought to you by..." or you can have them underwrite a specific program and use the appropriate language.
but the FCc has not hesitated to go after lpfm's for airing what sounds like a commercial, and you can get a mega fine.
so be careful out there. there are many folks who can look at your copy and let you know if it passes muster or toes the line. [*sic* throughout][44]

Again, this e-mail contained a level of detail uninteresting to a nonspecialist readership but spelled out rules that were very important for LPFM station holders to know; violation could result in the FCC imposing fines that would constitute a severe hardship for most LPFMs, which were often running on shoestring budgets.

In general, this list served as a resource for a variety of people and groups who had interest or involvement in LPFM. The tone and purpose of the e-mail list was one of sharing and it was expected that a range of people with expertise would offer advice and opinions on the variety of issues that came up, but Prometheus would step in to offer an authoritative answer when needed. Yet the list was plainly not only a place for "technical" information sharing but also a place to perform and reinforce the values and identity shared (and sometimes contested) by people who valued LPFM. In response to a concern someone raised about FCC penalties for a signal that was broadcasting "dead air", one list member posted: "Actually, I believe the LPFM reg's require a minimum amount of programming per week. Not sure if dead air would be considered 'programming'. Although it would still be better than the crap Clear Channel is airing."[45] This comment was a typical in-group display, reminding list members that they were united in opposition to the perceived antithesis of community media—consolidated, commercial Clear Channel.

Prometheus also maintained an e-mail list for more explicitly technical, purposes (related to hardware and software), called Leti. Their description of Leti read:

> Leti is the tech list for Prometheus Radio Project. We talk about audio, radio, and computer technology issues as they come up for community radio stations, and we use this list to plan for radio barnraisings. Leti is named for the comandante in charge of the famous El Salvadorian [*sic*] rebel radio station—Radio Benceremos.[46]

Leti—its title again signaling Prometheus's political leanings—had much less traffic and many fewer subscribers and posters than did Stubblefield. As the description indicates, it would see a flurry of traffic related to planning an upcoming barnraising, and it would often fall silent in between barnraisings. E-mails could discuss nitty-gritty ideas about how to lay out a studio:

> hey all, i have been reading the discussion about the studio layout from today, and decided i should send along some studio setups that i have seen. some ideas for studio layout, in general terms:
>
> —no through traffic, just one door if possible.
> —the layout should work for a solo programmer as well as for an engineer, interviewer and one to two guests. At least one surface should be kept clear between

the interviewer's and guests' microphones for glasses of water, microphone stands, books, scripts and other reference material.
—people don't like sitting with their back to the door.
—people entering the studio should be able to see the "On Air" light.
—the room doesn't need to be big, if the walls and corners are used well.
—the computer should be at a comfortable position to type at.
—turntables can be on a narrower shelf to the side.
—stand-up racks can be used for CD players and other audio equipment in corners.
—some people like to stand when on air, others like to sit—try to design so either type of programmer is comfortable.[47]

Or they might describe at length and in detail different options for handling the studio-transmitter link (a decision required of all new radio stations), including configuration, software, and operating system:

in the future you can send questions like this to the tech list—leti, and we can answer them there. for now, let me see if i can address your question.

i believe that philly imc page doesn't work any more,[48] but the software that the article refers to is called flow. flow was written in perl, and runs on linux. what it does is monitor streaming programs to make sure they are running correctly ...

([rolf, simon, liz]—if i am forgetting anything, or haven't represented flow correctly, let us know!)

the streaming link is routed over the internet, so a stable internet connection at each end is required. due to the fact that this is a streaming link over the internet, a delay of a few seconds (in some cases, 5+ seconds) because of encoding and transfer time over the internet will be encountered. this makes it impossible for the dj to use the air feed as a monitor, but that is usually not a big problem.

for w[xyz]-lp, at the florence barnraising, a point-to-point wifi link was set up between the streaming computer at the studio end, and the playback computer at the transmitter end. this has the effect of somewhat reducing the delay of the audio, and increasing the stability of the link, since the audio isn't routed over the public internet (and therefore out of your control) ...

there are other options, if linux is not a good choice for your staff....[49]

i do think that your best option is to have the antenna and transmitter located at the studio site. it vastly reduces the complexity of the setup, and allows for local monitoring of the air signal and transmitter status. the money you would spend on an engineer to check out your roof is worth it in not having to buy two more computers, and not having the link fail.[50]

These examples of the discussions on Leti indicate the content and tone of the list. Both of the e-mails I cite here were written by Prometheus's technical organizer, Brian, and posted to the list. But others besides Prometheus staff or organizers issued directives and advice. Similar to Stubblefield, Leti was intended as a participatory resource. There were many instances where Prometheus organizers asked technical questions to the lists and other list members gave authoritative answers. The information exchanged on this e-mail list was relatively detailed and the culture of the list was relatively expert.

Yet unlike many technical cultures, including the technicians described by Julian Orr, the tone of Leti did not tend toward competitive displays. Orr writes, "[Sometimes] the telling of a problem or a fix is a challenge. Do you understand what I'm telling you? Have you heard of this before? Do you recognize the symptoms?"[51] The sharing of highly specialized concerns can be accompanied by braggadocio, but the "duels" Orr witnessed among Xerox technicians did not feature on Leti. To the contrary, the language of Leti e-mails shows a collaborative environment in which expert knowledge could be shared, discussed, and respectfully and gently disputed. Previously, Brian wrote, "if i am forgetting anything, or haven't represented flow correctly, let [me] know!"; another e-mail from an Illinois-based Leti subscriber about where to locate a radio tower reads, "Before getting deeply into this, I wanta lay out the pro's and con's for different tower locations, as I see them. I'd greatly appreciate feedback on any misconceptions I might have on these, and/or on additional factors that need to be considered."[52] This language was not just lip service in a relatively expert community or false modesty. It was a deliberate effort to foster a civilized and inclusive environment (as well as to make sure that technical details were correct).

I witnessed only one dispute on Leti, about whether a new station would also stream or webcast its audio in addition to broadcasting on FM (chapter 7 addresses webcasting in greater detail). The language was fairly respectful and did not resemble a "flame war" that is typical in the e-mail list realm (and which I did witness on other e-mail groups devoted community radio). This could be in part because many of the people weighing in were not strangers to one another; I was reading the exchange in Philadelphia, but the majority of the participants were in the Illinois town where the new station would soon be built and knew each other personally. (Though this reason for the respectful tone is debatable, because vicious exchanges in online environments between people who know each other are certainly not uncommon.) But crucially, the most vehemently held opinions, which were expressed in the strongest language, were over what community radio or LPFM "is" or "should be." These opinions included the technical and licensing issues that were at stake, but focused mainly on the notion of "community" and ideas about what a community radio station's role ideally would be:

With all due respect, W[LMN] [the new station] is a station for [Holly] [the municipality], not for the next town over. As far as I am concerned, we are concerned, primarily, with providing space for self-expression by members of our community. Some see community radio as more concerned with providing content to listeners, and in this case streaming to the world can be seen as a benefit. For those who want a station to be firmly anchored in its community—either in terms of content produced, or in being a space in which all listeners can aspire to becoming an active participant, streaming becomes more problematic.[53]

Another Leti member then replied:

I am frankly STUPIFIED [*sic*] at the opposition to streaming ... "[WLMN] is to serve the [Holly] community." I completely disagree. What is the "[Holly]" community? At the VERY least, we are trying to serve the communities of [Holly, Laurel, Ashland, and University], if not the whole county. Am I mistaken? ... We don't even know if the on-air signal will reach downtown Laurel, let alone further. Hell, we can't even say if it will get to all of Holly.[54]

(I discuss the underpinnings of this debate and the issue of selecting which technologies to use to best serve the mission of community media in chapter 7, so I elide those matters here.) What is notable here is that the heatedness of this exchange was not a contest of technical virtuosity. It instead pertained to the mission of community radio generally and this station in particular.

The final aspect of Prometheus's technical work that I wish to describe I can only sketch, because during my fieldwork the organizers had only just gotten it systematically underway. They intended to codify much of the expertise that was shared informally among stations and more publicly on the e-mail lists and in workshops at barnraisings in the form of handbooks for stations. The plan was to have stations submit documents detailing how they dealt with a number of aspects of running their stations. Prometheus intended to archive this material and make available on their website, as well as cull it into handbooks. An intern wrote to Stubblefield soliciting the following materials:

Volunteer handbooks
Technical manuals
Fund-raising/underwriting documents
Programming guidelines
Governance documents
Bylaws

Mission statements
Station rules
Budgets
Training materials
Outreach materials
FCC communications
Operators logs/playlists/other things that keep your programmers organized[55]

As with the e-mail lists, Prometheus was intent on serving as a clearinghouse for information and a resource for LPFMs. They were purveyors of useful technical information and offered opinions and advice on a range of technical matters. Yet they were also trying to facilitate the exchange of this information directly between LPFMs in order to get people involved with LPFM to see *each other* (not only Prometheus) as a resource. Fundamentally, this was an organizing tactic, because Prometheus hoped to persuade LPFMs to see each other and themselves as members of a movement. Prometheus organizers sought to seed LPFM, diffusing both artifact and favored interpretations of small-scale broadcasting, and then they would step back. The next step was to let community members, station holders, and the artifact itself be responsible for the further growth and development of LPFM. In this, the radio activists' role as propagators of technology is evident.

The reader will note that I have largely left out descriptions of work with hardware and software being performed. This is mainly in the interest of not being repetitive. Barnraisings, Geek Group, and the transmitter-repair workshop discussed in preceding chapters also certainly "count" as technical activities, and would obviously be relevant to an analysis of the technical work in radio activism. But the reader will also note that technical *discussions* are highlighted in this chapter. This is a deliberate choice; I wish to accentuate the amount of language, knowledge, and discourse that circulated around technical practice in the universe of these workers. Technical practice includes not only the direct engagement of technical artifacts (as in hands-on tinkering) but also the enunciation of technical practice: what does it mean? what are best practices, and why?

Organizational Maturation and Boundary Drawing

Having laid out the activities that occupied the Prometheus workers, I now turn to a striking feature of the climate at Prometheus during my fieldwork: the group's struggle with its own success and its members' disquietude over maturing as a nonprofit organization. This played out in a variety ways. First, Prometheus workers took great pains to differentiate themselves from "mainstream" nonprofits, even those with who shared Prometheus's activist goals. Instead, they saw themselves (and wanted to be seen as) an "activist" group. Though I necessarily have a particular "snapshot" view of their work over time, I speculate that the construction and maintenance of this boundary was a relatively recent development. When Prometheus was younger, its members may have been more concerned with securing standing as legitimate spokespeople in legislative and legal situations. But during my time with the group, they saw themselves as poised on a threshold between gaining legitimacy on the one hand and remaining agitators on the other.[56]

Many illuminating moments occurred during a trip to Washington, DC, with Prometheus. At the second-ever meeting of Prometheus's board, held in an apartment, the group discussed challenges they faced in their work. Ellen told the board that Prometheus's work was harder than that of other groups that did media reform, because in her words, "Unlike Free Press, we don't carpetbag organize. We have to pay attention to the grassroots." She argued that Prometheus had to listen to "the voices of the people" in charting a course for the future of the organization. But justifying attention to spectrum management issues was particularly tough, because grassroots demand was low (see chapter 7 for more on this topic). Ellen's voicing of concern here can be read as identity work: she was uneasy about making the decision to pursue a policy agenda without a grassroots mandate because it would signal that the group was polluting its identity as "Washington outsiders." (Note also the "definition by opposition" role that Prometheus again placed Free Press in.)

Another instance of this tension over mainstream status was evident at the LPFM forum held at the FCC in 2005. These photographs illustrate the change in the FCC's stance toward low power radio: in the first, outside the FCC building in 1998, activists are seen launching a raucous protest (including broadcasting illegally into the building with a portable transmitter) (see figure 5.3). In the second image, the agitators have been invited inside the FCC (LPFM Day, previously described) (see figure 5.4). During the preparation for the LPFM forum, it was evident that Jasper felt a tinge of regret that he was making his point legally and largely on terms set by the FCC. He referred to the event as a "dog-and-pony show."

There was a real tension for Prometheus organizers over becoming too much of a "Beltway" organization (meaning a Washington, DC, "insider" organization). They expressed concern at the board meeting they were being forced to devote more resources to grant writing and policy work, as a result of their change in profile following their lawsuit against the

Figure 5.3
A rally in support of microradio in front of the Federal Communications Commission (1998). The switch is flipped to "turn on" free speech. Courtesy Mike Flugennock, http://www.sinkers.org.

FCC. Organizers requested that the board forcefully remind them of their mission to work with community groups. Back at home in the office in Philadelphia, they routinely teased each other for using words such as *constituents* and *deliverables* when talking about their work. Although I heard them use these terms, they were often peppered with comments such as "Oh, I can't believe I just said 'deliverables'—that's so horrible!" Again, these are displays of identity as the organization grappled with change. The organizers set up a binary between "Beltway" and "grassroots" and invoked the latter as the purest instantiation of their organizational priorities.

Ellen perfectly expressed the tension over this boundary when she described having bought a pair of "Congress pants" in a thrift store. These were pants that she felt looked like part of the "uniform" worn by lobbyists and other Washington insiders. She only wore them to dress up for occasions when she had to meet with members of this "other" group. She had clothes that made her feel like she "fit in" with the members of these other groups, but they were a "costume." She was proud of the fact that she had

Figure 5.4
Contrasting with the 1990s microradio era, in the 2000s, LPFM advocates were inside the regulatory process. A sign inside the FCC directing visitors to the Low Power FM Forum (February 8, 2005). Author photo.

not paid full retail price for them, in fact having bought them secondhand. Ellen also made a notable statement in a post to the office e-mail list. Prometheus had been asked to be a part of a coalition of advocacy groups working on media and democracy issues. Preceding the campaign director's visit to Philadelphia, Ellen wrote to other staff, "so, it would be nice to impress the campaign director of this coalition of doom and also pain, but it isn't the most important thing on my mind. [A]t any rate, he's coming up on [F]riday to meet with [P]hiladelphia organizations, so we should suggest some he should meet with."[57] It is obvious from her language ("coalition of doom and also pain") and her blasé attitude about the need to impress the director that she was not thrilled about working with (or identifying with) what she perceived as an unwieldy coalition, some of whose other members were more "mainstream" advocacy nonprofit groups.

On a road trip with organizers, it took me a little while to realize that the decision to stay on people's floors as we drove cross-country was not merely due to financial considerations. It was an active part of Prometheus's organizing strategy, intended to make a point to people that activists wanted to connect with them in their homes and that they would accept lodging and food offered by community members. There was an implicit statement that they supported a vision of a communal, in-kind, reciprocal economy in which labor, housing, services, and food were provided freely in response to needs. This served to create a sense of having a tangible, reciprocal relationship with the grassroots. It also differentiated Prometheus from mainstream professionals (including those at other nonprofits), who work and travel quite differently.

During a job search, Prometheus further displayed their organizational commitments. The full-time paid staff increased from three to four people during my fieldwork. The organization searched for a staff member to oversee technical services (including working closely with station applicants and license-holders). During this search, Jasper and Ellen had a minor dispute over the job description Jasper had drafted. In it, he had listed "Must not be afraid of mice, our office sometimes has a lot of them"[58] in the list of attributes desired in a candidate. Ellen was dismayed by this, asking him, "How can you expect anyone to take us seriously if we don't take ourselves seriously?" Other interns and volunteers thought this description was funny, not inaccurate, and okay to leave in. Eventually they reached agreement that she would rewrite the description. She removed the language about mice, but left intact much of Jasper's draft, including, "Dress is casual, though on occasion you may need to bathe and pull on something that has no holes in it."[59] This indicates that this part of the ad was uncontroversial in its representation of the group's identity.

The hiring stage of the search was illuminating, too. After interviews, the staff had narrowed the pool down three or four qualified candidates, each with strengths and weaknesses. It was agreed that two of the applicants (Brian and a woman named Robin) were very strong with attributes that complemented each other. The staff decided to offer the job to both of them as two half-time positions. Ultimately, Robin found a different job and turned them down, resulting in Brian being hired to fill the position full-time. In itself, this is unremarkable. But Robin's politics were so fully radicalized that she did not believe in government regulation of the airwaves; she had acknowledged that she would not deal with the FCC if she took the job. In spite of this, the Prometheus staff was fully willing to hire her as long as they also hired Brian to cover the parts of the work that would require dealing with the FCC. Perhaps because Robin's beliefs represented an idealized notion of Prometheus's mission, the group was willing to reconfigure the job description in order to hire her. This was the case even though filling one job with two people would clearly be more trouble.

What is evident here is activist discomfort over becoming a mainstream nonprofit. This was manifest in myriad ways; it was also a subset of a more general performance of criticism of mainstream values. In particular, "mainstream nonprofits" were viewed with suspicion. Activists saw them as being too close to establishment values and as not being radicalized enough. These "other" organizations also represented a threat due to their potential to change activists into "insiders" who have lost sight of their activism.[60] The point is not that these distinctions between groups were necessarily "true." Instead, we should understand them as folklore. Drawing contrasts between themselves and "mainstream" nonprofits reassured Prometheus workers that they were scrappy agitators with a connection to the grassroots. Such distinctions were of great importance to the organization as it sought to define itself and chart a course forward.

Labor, Craft, and Technology: Drawing Boundaries Using Technical Affinity

Anxiety over organizational change is common. Activists (including radio activists) often see their radical stances attenuate over time. What stands out here is that the radio activists invoked technical work in a novel way. They used it as an ideational resource to mark differences between themselves and "other" groups, especially "mainstream nonprofits" and "policy insiders." This strategy of drawing boundaries and managing organizational change seems uniquely adapted to technological activism.

Work with machines has been theorized as a common site for identity formation. Most elementally, "technical identity" is usually understood as a cause and a consequence of participation in technical occupations and hobbies. Technical identity also conveys a closer relationship with technology than that held by average users and is usually constituted in part by an affective relationship with technology.[61] This book builds on this useful but rather narrow starting point for understanding technical identity to examine it as a resource for a group whose work is not primarily technical. This breadth of tasks (including paperwork and organizing work) differentiates the radio activists from hackers, radio hams, or machine technicians.[62] The radio activists were concerned with producing working radio hardware and desirable social relations through the democratization of expertise. Their work was not limited to technical engagement. They also labored in advocacy around policy (as previously discussed) and in significant amounts of paperwork, filing, and other office and organizational work (also previously described). Here, technical identity extends beyond technical engagement itself. It served an important function within a wider repertoire of activist practices and commitments.

Having laid out how technical identity is constructed in chapters 2 and 3, I give that topic lighter treatment here. In the following section, I focus mainly on how Prometheus's technical identity was made manifest when they interacted with "outsiders." These included members of the public, other nonprofits, and Washington "insiders." One simple distillation of this is t-shirts that Prometheans wore and sold (at barnraisings and on their website). The t-shirt showed a schematic for a transmitter (see figure 5.5). To wear this schematic was to display a symbol of the organization and its core values.

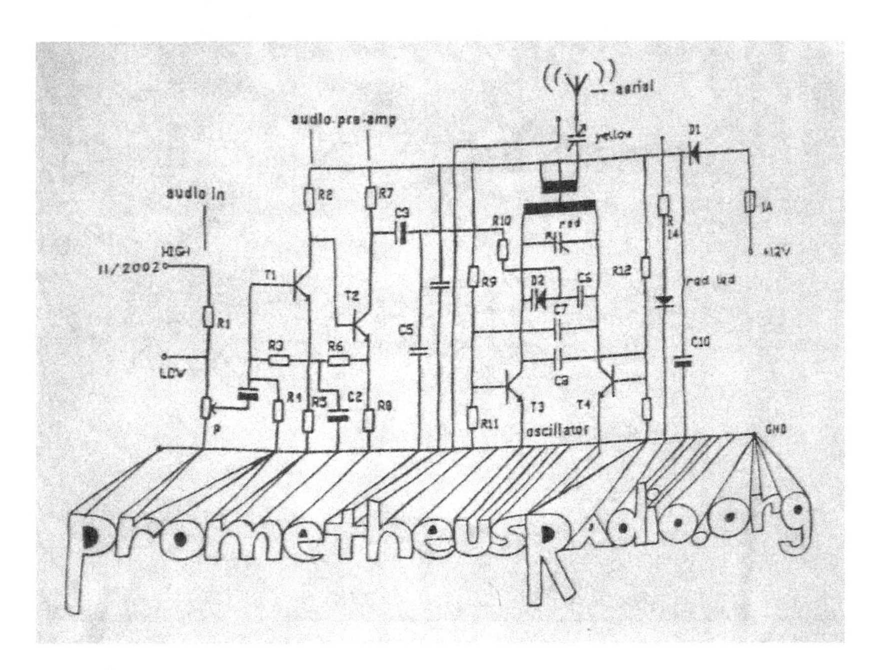

Figure 5.5
The back of a Prometheus Radio Project t-shirt (ca. 2005). Author photo.

It is important to recall that technical or geek *identity* is not synonymous technical *skill*. Not everyone in the group was equally skilled with radio hardware or computers. But the deployment of geek identity was a resource on which the workers routinely drew. It is worth disaggregating technical expertise from a technical identity. People with relatively less expertise may identify relatively strongly as geeks. This is partly because geek identity here was constructed in a manner that self-consciously valued participatory ideals, as opposed to holding technology (and technical identity) as the exclusive domain of experts. An explicit goal of the organization was to "demystify" technology.

One noteworthy performance of technical identity was in Louisville, Kentucky, at a rally for the Coalition of Immokalee Workers (a group of Mexican, Guatemalan, and Haitian migrant workers with whom Prometheus built a radio station, *Radio Consciencia*, in 2003). It was a windy day and Prometheus's literature kept threatening to blow off a table on which flyers and books were arrayed. Jasper and I ran to the car to get suitable paperweights, and he grabbed a box of tools.

Figure 5.6
Tools holding down literature, Louisville, Kentucky (March 2005). Author photo.

I noticed that many people who approached the table complimented the tools (see figure 5.6), comments such as "Hey, nice paperweights!"

The radio activists' use of carpentry and electronics tools for paperweights was freighted with symbolism. The tools served to display technical identity. But there were additional layers of meaning. The rally was in celebration of a major labor victory by the Immokalee workers against the corporate parent of Taco Bell, for whom they picked tomatoes, and some of the people who complimented the tools were wearing pro-union t-shirts.[63] In the simple act of weighting down their literature, Prometheus performed solidarity with craftspeople and showed their belief in a participatory and hands-on relationship with technology. Both of these were important markers of identity for their group within the terrain of media democracy work.

This ties into that the issue of labor politics and identity. Prometheus routinely positioned themselves as "not-professionals." Jasper said in the aforementioned board meeting that they did not have a professional standing, "we are not engineers or lawyers."[64] He also stressed the importance of giving presentations that were "comprehensible to humans." He made an effort to distance himself from the expertise held by engineers and lawyers, advising the other staff that

> when they open their mouths, pure gobbledygook comes out that is incomprehensible to humans ... you can [be inaccurate about some details] when you talk in public, you need to just be evocative about the scale of things. lawyers and engineers are professionally responsible to be precise and nitpicky and boring ... leave the lawyer/engineer details for the handouts and the personal conversations.[65]

He also maintained that when he began his work traveling around the country teaching people to fill out LPFM license applications in 2000, it was not what he had expected to do with his life. He still identified strongly with his past as a pirate. He also stated at this board meeting that he needed guidance from the board on how to run the organization, because he "hated nonprofits" and was "used to hanging doors for a living" (referring to his earlier work as an unlicensed "black-market carpenter," and again invoking tools).[66] Another organizer, Brian, was employed as a card-counting professional gambler in Las Vegas in between his punk days and work for Prometheus. In sum, although craft and technique were important to these workers, professionalization per se was not. "Anti-professionalization" was a hallmark of their self-understandings.

Thomas further underscored this when he said in an interview, "I'm just a bearded guy in a basement," indicating that the organization's work was conducted in a setting that might not seem like a traditional office. In February

2006, Brian brought up the issue of unionizing the office during discussions about obtaining health insurance. He professed his intent to join the Wobblies (Industrial Workers of the World), stating that it might be useful in the fee-for-service projects when they were working alongside tradespeople.[67]

Conclusion: Drawing and Effacing Boundaries with Technical Identity

Prometheus's work was shot through with technical identity. But notably, it was invoked in a counterintuitive manner: it was used to reinforce an identity that was non-professional, antitechnocratic, and not white collar.[68] What is the explanation for this iteration of technical identity? It is curious given that these activists worked in an office.

One answer is that Prometheus not only negotiated boundaries vis-à-vis "mainstream" nonprofits and tradespeople, but also vis-à-vis other members of the microradio community whose politics were more radicalized.[69] In a 2005 staff meeting discussing coordination for an upcoming conference on community radio, activists talked about the potential for conflict between famed Bay Area pirate activist Stephen Dunifer and FCC members. They were unsure whether Dunifer would "be nasty to the FCC, or encourage other people to be nasty to them." They were also concerned that if Dunifer wanted to lead transmitter-building workshops in which he would encourage the transmitters to be used for unlicensed broadcasting, the FCC might be unappreciative.[70] (Ellen said that one solution would be to just say that the use of the transmitters internationally, outside of the FCC's domain, and likened the situation to buying a pipe for smoking marijuana: "Have you ever been to a head shop? You need to say, 'I'd like to see your water pipes, please.' Just watch the language and it'll be okay."[71]) Ultimately, Dunifer did not attend the conference, but the head of the FCC's audio division spoke on a panel (Thomas had joked about putting them on the same panel). Prometheus's position between "pirates" (whom they were not sure would "play nice") and the FCC illustrates a situation in which they felt they needed to reflect on these differences and their own role.

In practice, radio activists performed a balancing act between oppositional and institutional politics. As former pirates clamoring for increased and better regulation in their policy work, they embodied contradiction. As one activist said, "I'm working to advance goals within liberal reforms that are consonant with a more radical vision. I have radical ideals, but I'm also a pragmatist."[72] There is stark contrast here with other strains of techno-libertarian engagement, including some hackers and free software communities, as well as more radicalized or anarchist proponents of microradio.

The assertion of technical and craft identity under these circumstances was a way of managing anxiety about the attenuation of radicalism in the face of a more professionalized, more expert work environment. This identity served to establish continuity and efface boundaries between the organization's past, present, and future. This technical identity is suffused and conjoined with political radicalism, in spite of the deep contradictions between the elite nature of technical expertise and the ethos of radical participation the activists valued. Instead, the organizers' emphasis on technical or geek identity was a way of performing a "we're not them" position vis-à-vis a more professionalized, insider, and expert identity. They wished to define themselves in contrast to the "others" they encountered in nonprofit and policy settings.

Ironically, this iteration of technical identity still carried its own exclusiveness, even though it was constructed as accessible to members of the public and was intended to bolster a sense of unity among activists. Technologically oriented activists may pride themselves on their geekiness and their failure to cultivate the slicker interpersonal style of lobbyists and of political insiders.[73] These examples—from flashing at a rally, to wearing "Congress pants," to their sometimes uneasy position between pirates and the FCC—demonstrate that the radio activists made significant efforts to resist embodying the role of Beltway insiders. They prided themselves on doing what lobbyists and political insiders ostensibly did not: getting their hands dirty, actually building something. To drive this point home, at a 2005 "birthday party" for LPFM in Washington (celebrating the five-year anniversary of the FCC's designation of LPFM service), Prometheus presented their lawyer with a handmade radio receiver from South Africa. This was not only an expression of gratitude. It was also a way of demonstrating that the lawyer's "difference" was overcome, thus effacing the boundary they policed when they valorized technical engagement.[74]

This expressive culture did important work for the radio activists. They sought to mature as an organization and refine their mission while continuing to work in a manner perceived to be in line with radical activist ideals. Their valorization of technical affinity and craft identity romanticized an outsider identity vis-à-vis the political insiders with whom they were increasingly interacting. On close inspection, these forms of oppositional display bore much in common with piratical ideals. They marked the radio activists as outside of professionalism, as self-reliant, and as technically savvy (recall the activist claim that, borne of necessity, "pirates have the best tech skills"). Of course, these distinctions between groups are not necessarily "true."[75] The elevation of the technical and craft elements of their work meant that they systematically downplayed their actual organizational bureaucracy, their increasing policy expertise, and their educational attainment (a majority of the full-time staff held degrees from elite private universities, for example). Nonetheless, this strategy seems uniquely adapted to the pirate history and technical commitments of this group: other groups not engaged in a technological form of activism may find different strategies to manage the maintenance of an activist identity. For Prometheus, technical affinity helped to maintain the idea of a boundary between what these workers believed their own roles to be in relation to members of "mainstream nonprofit" and "policy" occupations.

And yet, in practice, much of the radio activists' work during this period was mundane office work. This returns us to a discussion of what activism is. It also raises the issue of how the group was able to perceive its activities as unified. As previously outlined, staff tasks and work practices were varied, but the organizers nonetheless experienced coherence. Their job titles illustrate the fluidity and equivalence in their positions; even flippant variability in titles did not translate into inconsistency for the workers.[76] The collaborative nature of the work in the office also bolstered the sense that everyone did the same work. Though paid workers did have different domains, they all collaborated on projects in each other's areas at times. They all shared (or delegated to volunteers and interns) grunt work such as photocopying, and all office members (paid and unpaid) shared cleaning and office maintenance. There was no one whose duties were purely "secretarial," "administrative," or "technical." Another contributing factor in the coherence of their occupational identity may be that the Prometheus workers saw their work as *productive* activity (as opposed to *paid* activity), as discussed in chapter 4.

Organizers emphasized repeatedly that they would do the work of activism regardless, and were lucky to be paid for it. Yet Prometheus was becoming attentive to the issue of making the organization's sustainability possible and even routine. Thomas pointed to the perceived difference of Prometheus's past from present when he said in an interview that "five years ago, our organizational strategy wasn't 'How can we make this organization sustainable [through grant writing, getting health insurance for workers, etc.]?' It was 'What can we chain ourselves to to make the greatest impact? What is the most effective thing we can chain ourselves to?'"[77] This statement, though an exaggerated version of reality, is telling because it reveals the purported commitment of Prometheus workers to activist goals (and direct-action strategies) in spite of the organization's trajectory as a sustainable nonprofit. It also characterizes the "work" of Prometheus as direct action and other classic activist strategies.

Revisiting Thomas's statement, one notes that he did not say that a past organizational strategy was "building pirate radio stations." He said, "chaining ourselves to things." This raises a final point about what activism *is*. For Prometheus workers, there was constant slippage with regard to their primary focus. As noted in chapter 4, Ellen's e-mail signature read "building radio stations = awesome."[78] This might indicate that the main objective of these actors is to build community stations (certainly, this e-mail signature can be read as a performance that indicates this work is of great significance or even primary importance). And yet Thomas claimed that building radio stations was really "just sleight of hand" for their real work of building a movement and critiquing power relations.

This point becomes additionally clear in Thomas's characterization of paperwork as not being a hindrance to their main mission. To the contrary, the paperwork was becoming a part of "good activism" because it enabled them to sustain and expand the activities of the organization. Thomas said that a lot of the administrative work Prometheus encountered during my fieldwork was time-consuming for him at that transitional moment, but he hoped it would require much less work once he got the hang of it. It was very challenging to research and acquire insurance, to learn to do proper accounting, and to meet with consultants on both of these issues. But he felt that these were essentially one-time propositions. He hoped that after he learned how to maintain these systems, he would spend less time on them.

Thomas's statement that administrative work did not pose a hindrance to Prometheus's work stands in contradiction to the anxiety and exasperation expressed at other times by other organizers. Recalling Jasper's comment in the board meeting, sometimes the activists expressed dismay that administrative work (in particular grant writing or "chasing money") could hinder their ability to allocate time and energy toward other aspects of work such as organizing or technical work.[79] It seems that one strategy for the organizers was to view administrative work as part of the work of activism. In doing so, they reclaimed this labor. They refused to view their tasks as stratified, and vested even tiresome bureaucratic work with importance and consonance.

Rather than accepting any of these claims at face value, I posit that activist identity is the best way to understand what activism *is*. These actors drew heavily on a shared sense of activist identity in order to draw boundaries between themselves and "other" actors in the realm of media democracy work (primarily "mainstream" nonprofits and Washington "insiders"). Technical affinity and craft identity largely bolstered activist identity. Deploying the two together helped them to manage the issue of "outsiders" and to efface potential boundaries between "insiders." A sense of activist identity additionally effaced boundaries within the group and enhanced the sense that the work performed by the group was coherent, even though the tasks varied and the priorities of the organization shifted. Distinctions between work tasks did not worry them. In fact, varying practices were assigned coherence by framing them as expressions of activist agency. If it is not too tautological, it seems that the best working definition of activism can be found in the actors' self-understanding of their work as activism and themselves as activists. The role of technical identity within activism is particularly important for how it may be mobilized to shore up activist ideals.

6 Making Old Technology Anew: Reinventing FM Radio in the Twenty-First Century

In 2006, Prometheus held a barnraising with a farmworkers' union in Oregon. Founded in the 1980s, the union represented tree planters and farmworkers, and was Oregon's largest Latino organization.[1] Throughout the barnraising weekend, activists and union organizers stressed the potential for the new station to extend the union's organizing activities. At a plenary session, an eloquent and powerful audio greeting from an activist in Peru intoned:

> Dear friends, [simply broadcasting] information is not enough. We need to position ourselves, we have to get involved, we need to strive to turn community radio [station]s into actors—not only spectators, but actors in society; not only informing others what is happening, but transforming the reality.... Here we say and believe that a different world is possible.[2] We also say and believe that a different radio is possible.[3]

By the time of this barnraising, radio broadcasting was nearly ninety years old. Yet the meaning of radio broadcasting has not been constant across time. In the Peruvian activist's statement, we can hear efforts to define its meaning in the twenty-first century. We can also hear contestation over its use. The Peruvian activist stressed radio as an *actor* in society—not as a conduit for "informing," but a tool for transformation. In his claim that "a different radio is possible," he challenged uses of radio that would exclude this transformative, political component.

What meanings did radio activists impute to radio, and how did their interpretations deviate from dominant meanings of broadcasting? How can we gauge these interpretations? And why do these questions matter—for media activism or for social studies of technology? This chapter considers radio activists' recent efforts to shape radio technology and associated practices. Radio provides a unique opportunity to view the flexibility of technology over time—in this case, decades after being introduced as a "new technology." Radio has faced recurrent debates over its meaning and use, making it especially rich territory for understanding the social shaping of technology over time.[4] Writing in 1947, producer-turned-critic Albert N. Williams argued that "radio is not, today, any of the things it was born to be. It is not operated in the public interest. It is operated in the specific interest of [corporate sponsors of programming].... It is only an educational, political, and social force after the salesmen have enjoyed their sport."[5] Williams's critique appeared nearly twenty years after the national network system of advertising-funded radio had stabilized. But his comments could have been made by the Prometheus activists another fifty years later. In thinking about longitudinal change, we should understand radio broadcasting as situated. Its prior contested history provides insight into the sedimented meanings it had accrued by the 1990s, the period in which the activists encountered and interpreted it. At the same time, the recent phenomenon of radio activism must be viewed in light of contemporary social, technical, and political arrangements. The activists challenged the dominant meaning of broadcasting and the social arrangements surrounding it.[6]

This chapter examines radio activists' preferred interpretations of LPFM. I turn away from technical work and toward advocacy and organizing in order to demonstrate how activists attempted to convince members of other groups (including community groups and regulators) that certain uses of low power radio technology were most appropriate. They specifically emphasized radio's significance for political organizing, for the promotion of "localism," and for small-scale noncommercial use. (Needless to say, these are not discrete rubrics, but I tease them apart as descriptive categories in order to highlight the most prized interpretations of radio for the activists.) Significantly, these interpretations were not shared by some other social groups, and thus the radio activists entered into an agonistic field in which some interpretations triumphed over others in terms of public perception and in terms of inscription into policy.

The radio activists occupied a mediating position, situated between members of the public and regulators. They challenged dominant interpretations of FM radio in order to make a case for its contemporary value and viability as a communication medium. Focusing on the period from 1998 to 2006 (which includes the period before LPFM was introduced by the FCC and the early years of the LPFM service), I outline the activists' efforts to promote their interpretations of LPFM. A contestation over the appropriate meaning and use of FM radio technology was at the heart of the activists' undertakings. In addition to mediating between groups, their ultimate goal was to propagate the technology itself. Although they largely believed that their interpretations constituted appropriate use of the technology, their actions were not limited to interpretive, mediating work; as the preceding chapters show, Prometheus also worked directly to build and distribute broadcast technology. This intersection of technical practice and political engagement differentiates technological activism from other sorts of advocacy as well as other technical pursuits. This makes propagators a distinct group within the category of mediators.

In the following sections, I outline the interpretations articulated by the Prometheus activists and elaborate how members of different groups agreed or disagreed about these interpretations of LPFM. Though material and discursive practices are intertwined, this chapter highlights rhetorical work.[7]

Political Organizing

Regulators have long had an uneasy relationship with content. The Radio Act of 1927 "called for the [Federal Radio Commission (FRC), precursor to the modern-day FCC] to allocate licenses on the basis of which prospective broadcaster best served the 'public interest, convenience, or necessity.'" This language was borrowed directly from public utilities law.[8] The FRC chose to interpret "general" broadcasters supported through advertising revenue as those who served the public good. The FRC contrasted "general" broadcasting with "propaganda" stations held by groups dedicated to espousing particular viewpoints on the air.[9] The FRC also determined that because every "propaganda" viewpoint could not be represented on the air, "general" broadcasters were to be heavily favored over propaganda stations in the allocation of licenses. This dealt a severe blow to broadcasters who were funded in ways other than advertising. The FRC's refusal to privilege certain "viewpoints" flowed from the FRC's interpretation of its mandate. The FRC claimed it could not regulate content beyond very general parameters (including protecting the public from indecency) due to the First Amendment's prohibition on abridging free speech or press freedom. Yet the FRC "served less as an impartial arbiter than as a cynical handmaiden for network interests."[10] This crystallized early broadcasting as networked, corporate, and supported by advertising. Educational, not-for-profit, and independent stations suffered. In 1949, the FCC tried to redefine the proper use of the airwaves by saying that radio should serve the public interest. It was met with resistance from commercial broadcasters, who tarred the FCC as communists or fascists for attempting to define and enforce public interest guidelines.[11] In response, the FCC muted its recommendations to coalesce around the notion that stations airing programming of a controversial nature should be required to air the views of competing interests (known as the "Fairness Doctrine"). Much of the Fairness Doctrine was officially abandoned in 1987 under FCC Chairman Mark Fowler (a Reagan appointee).[12] Over time, the FCC has been reluctant to regulate broadcast content and has regarded commercial use as a value-neutral proposition.[13]

Producers of "radical" media, including LPFM and community radio, often view media as a crucial aspect of "helping give voice to oppositional politics and cultures."[14] The radio activists defined radio as a tool that could empower citizens and community groups to effect political change. In this section, I sketch examples of the activists and other groups (particularly community groups seeking radio stations) who identified political use as a main purpose of a radio station. In these examples, the activists mediated between groups to promote the idea that radio stations were political tools, particularly to community groups and the public.

Barnraisings were sites for political consciousness raising. For activists, the political potential of a new radio station was paramount. This was particularly evident at the 2006 barnraising for the farmworkers' union in Oregon. Prometheus was unambiguous about supporting this group because of its larger mission. Ellen explained, "To be totally blatant, one of the reasons we're so excited to do a barnraising with [the union] is that we hope that you guys [other LPFMs and volunteers in attendance] will take away this idea that the station is a tool we own for responding ... to issues of justice in communities—its power for organizing."[15]

A prominent issue at this barnraising was immigration, which was for obvious reasons a matter of central concern for this group of workers. In his greeting, the Peruvian activist commented on the centrality of media mobilization around the immigration issue. He indicated that the union's station could similarly be used to mobilize people:

> I was told, for example, that in this year's May Day marches, when millions of Mexicans, millions of Latino immigrants went out into the streets to protests, many radio stations accompanied the struggle, and participated together with the people. And your radio [station], where was *your* radio [station] that day? Playing a little music? I don't think so! Surely your radio [station] was accompanying the popular struggle.[16]

Summer 2006 was a heady time for this issue. Earlier that year, the issue of immigration had come to prominence in the mainstream media. Congress and the public gave much attention to immigration during the lead-up to the 2006 midterm elections. Many large-scale marches protesting the criminalization of undocumented immigrants occurred across the country. The Oregon station prioritized immigration as one of its many social justice goals. At a workshop entitled "Radio and Labor Organizing," volunteers from Immokalee, Florida, Woodburn, Oregon, and Philadelphia shared their experiences using media to organize in their local environments. Workshop leaders highlighted the use of media people to march and speak out for immigrants' rights. Mainstream media emphasized the role of Spanish-language media, including radio stations (notably KSCA in Los Angeles), in galvanizing supporters of immigration rights and amnesty.[17] A union member named Esteban said that "before the spring, 5,000 people would come to the marches. With the radio station, then there were 30,000. With a radio station, we can have 20,000 more people. This radio station will allow us to organize."[18]

Prometheus planned the barnraising so that representatives from another migrant farmworkers' union, the Florida-based Coalition of Immokalee Workers (CIW), could be in attendance to help run the event. Prometheus had held a barnraising in 2003 with the CIW. During that period, the CIW had organized a successful four-year boycott of Taco Bell and its parent corporation Yum! Brands over the wages paid to tomato pickers. They won this campaign in 2005 when the company agreed to a wage increase.[19] The CIW representatives shared how they used their radio station. Maria stated,

> We use radio because it's hard to get to all the sites, or in case we get to a home and no one answers the door.... Radio helped us do many things, tell people about their rights. It's good for people just arriving from other countries.... Many people in other countries believe the US is beautiful and marvelous and has lots of money, and they don't know about the exploitation [faced by migrants]. But we communicate about our experiences, we make connections with people in other countries, [and now] between CIW and [the Oregon union].[20]

Echoing her enthusiasm, Esteban commented on his vision for the Oregon station:

> The station will not just be for campesinos, but for the whole community. Youth will be able to speak and learn. Campesinos work more than twelve hours per day, they don't have TV, [and] sometimes there is no physical place [for union organizers] to visit, but radio will reach them. [Our goals for this station are to]: To educate people, to train them *not to listen to radio but to use it.*[21]

Notably, these station holders emphasized the political applications of their stations; Maria also spoke of a burgeoning connection between the two unions' stations, envisioning them as linked in a broad movement underpinned by communication.

At another workshop, Prometheus organizer Ellen described how the CIW used its radio station. After a hurricane in their region of Florida, some workers were hired by a nonlocal company contracted by the federal government to perform cleanup. After several weeks, a few people talking among themselves noted that no one had yet been paid for their work cleaning up trees and other debris. They went to the radio station and broadcasted a request that everyone who had not been paid gather that evening. When they were all assembled, they realized that fully two hundred workers had not been paid. They contacted the management of the company, which was based out of state, and a manager flew in the next day to issue checks. Prometheus activists told this story often, and it functioned as folklore for them. When Jasper re-told it, he added that it proved that "radio isn't just a jukebox."[22] Broadcasters and activists alike understood the radio station as a tool to raise consciousness about the similar situations of many individual workers who would be less aware of their collective plight without the radio station to draw attention to it. The CIW's station was particularly valuable for its ability to reach people without stable residences or who might not answer their doors (perhaps fearing visits from immigration authorities). It also had the ability to reach people who spoke a range of languages. Some of the CIW members natively spoke indigenous Central American languages; Spanish was their second language, and their literacy might be limited. The Oregon farmworkers pursued their station for identical reasons, highlighting linguistic differences and other voices marginalized in mainstream media: "Many [on our station] are voice[s] never heard on commercial Spanish-language radio or by a mass audience. They include *youth, indigenous people* (speaking in their own language), *women,* workers, plus *progressive leaders from Latino communities around the corner and around Latin America."*[23]

In 2005, Prometheus built another low-power community radio station near Arusha, Tanzania. They built it in cooperation with a community center run by American former Black Panthers who were living in exile. Over a series of days, the community center members discussed the mission for the station. One concern was that, unlike in the United States, the station would not have the freedom to openly discuss politics. They decided that the station would seek a license and legal status, which would give it a greater likelihood of longevity. This decision meant they would need to keep in good standing with the government. It would be preferable to avoid explicit coverage of some potentially controversial topics such as elections.[24] The station's volunteers wondered "how to be political without being overtly political."

A young man from South Africa, Safi, who was active in an unlicensed community station in Johannesburg (motto: "We mic what we live"), led a workshop to plan programming for the new station. Safi recommended that the station seek a wide range of volunteers. In an explicit critique of mainstream media and elitist notions of expertise, he said, "This is about getting as many people as possible involved in creating knowledge."[25] He and the other volunteers planned some potential programs, including a hip-hop[26] program, a program for primary-school-age children to get help with schoolwork, and a program in which they planned to interview community elders about various topics, tentatively titled "What's up, *Wazee* [elders]?"[27] The station also generated a list of rules for its operation, which I reproduce at length:

> NO SEXISM. This means that [we] will not play music that promotes the degradation, exploitation, or oppression of women. Songs whose theme is pimping women, have extreme references to physical or sexual abuse, or which generally promote sexist attitudes will not be played by any DJ.
> NO RACISM. [We are] about promoting community development and upliftment, specifically within the African context. We will not air any programming which promotes a racist ideology. This includes any song, show, or commentary that degrades African culture, aesthetic, and traditions. This does not apply to critical discussion, but is meant to prevent the promotion of values and standards that place all things African in an inferior position.
> PROMOTE COMMUNITY. [This station] is the voice of [our] Community Center. At all times the programming should reflect the goals and objectives of the [center], namely, "Sharing Knowledge for Community Development." This rule is meant to discourage programming that is individualistic and materialistic in nature. We should at all times promote the principles of *kujitegemea* [self-reliance], *kujichagulia* [self-determination], and *ujamaa* [cooperative economics, fellowship].[28]

These rules codified an explicit code of conduct for broadcasters that reflected the political ideals of the community center, including belief in moral uplift, change from within the community, and a combination of self-determination and fellowship. This was a thoughtful and overtly political use for the station, though it skirted electoral politics.

Broadcasters and radio activists alike viewed the radio stations as a means to extend and maximize the political and social work in which the organizations already participated. In the Tanzanian community center's case, the goals for the station's organizing potential were perhaps more diffuse and less direct than those of the US farmworker unions. In the words of the Peruvian activist who gave the address at the Oregon barnraising, the stations themselves were "actors in society; not only informing others what is happening, but transforming the reality."

The activists fundamentally conceived of radio as a tool for enacting social and political change. But there was, at times, an uneasy distinction between content and form. The activists often imagined radio stations as being conduits for content, and they were savvy about the desire to "stack the deck" by building media outlets whose politics were somewhat aligned with their own.[29] But they simultaneously believed that the technology of low-power radio *itself* was inherently transformative, perhaps even inherently democratizing. Radio stations in the hands of citizens would on their own produce social change. (This is type of rhetoric is very common, particularly during the introductory period of many new communication technologies. In more recent memory, the early years of the Internet saw it hailed as a technology with revolutionary attributes.[30])

This tension between content and form is illustrated in the following examples. In Jasper's view, a radio station was almost a microcosm of society. A station could contain a kernel of the social relations activists hoped for in the society at large. He said:

> [C]onflicts are very common at community radio stations. First, you should remember to think of it as positive that people care so much enough about your station that they are willing to argue with each other about it....
>
> [But i]f you think about it, it is remarkable how little people fight at community radio stations. After all, your mission is to bring together diverse constituencies from around your community and put them all together, almost all of them working as volunteers, to make the radio station work.... One of the greatest benefits of community stations is that they teach us how to administer a community project democratically, and help prepare people for a freer society than the one that we have.[31]

In essence, Jasper claimed that even an apparent disadvantage of radio—the scarcity of airtime—was a plus in that it prompted groups to make thoughtful decisions about how to use their stations. Often, this led stations to explicitly focus on community governance.[32] In connecting governance of a station to a "freer society," he demonstrated his deeply held normative understanding of the potential change wrought by community media.

Conversely, in a 2005 discussion with members of the House Democratic Progressive Caucus in the US Capitol building, Jasper disclosed that Prometheus's goal was "to get progressive radio not just in the big cities but into every nook and cranny around the country."[33] Representative Maurice Hinchey (D-NY) responded by saying that the country was witnessing the fruits of a "twenty-year right-wing plan to control information people receive, which began in 1987 when the Reagan administration did away with the Fairness Doctrine. If you can have a monopoly on information, you can control the political process."[34] Groups such as the activists and lawmakers were quite attuned to the issue of political content in broadcast media, although the boundary between artifact and content was sometimes blurred.

Localism

The radio activists placed a premium on media outlets with ties to their immediate communities. They held "localism" to be intrinsically meaningful in itself, independent of any "political" uses for stations.[35] LPFM advocates highlighted multiple rationales for the importance of localism. They suggested that local news and coverage of local affairs served a general public good. They also cited pluralism, the promotion and sustainability of musical artists, and public safety and disaster preparedness to be inherently beneficial uses of LPFM that were tied to localism. In all cases, local programming was contrasted with syndicated programming carried by commercial and noncommercial stations. (The radio activists themselves cited an interest in localism that emerged in part from radical politics, particularly anarcho-syndicalism—a lineage that was almost certainly not shared by other groups with whom they collaborated.[36])

The radio activists' elevation of localism significantly diverges from the enduring national, networked interpretation of broadcasting first championed by the Federal Radio Commission in the 1920s and 1930s. In the early days of broadcasting, the urge for radio to span distance was strong (partly due to the legacy of the practices of amateur operators in the 1910s and 1920s). In fact, one use of radio was to try to transmit and receive signals from as far away as possible, an exercise called "DXing," practiced by so-called distance fiends.[37] Some broadcasters and listeners wanted radio technology to conquer space for the novelty, pleasure, and curiosity of it.[38] This idea of listening across distance contributed to the formation of an imagined national audience. For the first time, many types of music and sports programming were popular across regions and audiences. Reformers and educators, including John Dewey and Robert Park, "construed modern communication essentially as an agent for restoring a broad moral and political consensus to America, a consensus they believed to have been threatened by the wrenching disruptions of the nineteenth century: industrialization, urbanization, and immigration."[39] The conception of a national audience created the mandate to make nationwide broadcasting technologically possible. Technological options available to the radio industry included "superpower" broadcast (AM), shortwave, and wired networks.

The wired network became the option of choice, resulting from the interplay between radio giants such as the Radio Corporation of America (RCA) and AT&T, which had a firm hold on telephony. The antitrust atmosphere of the 1920s made these companies highly wary of being perceived as monopolistic. Ironically, though, the patent negotiations between the companies resulted in RCA being largely responsible for setting up and administering the wired network system (even though local stations receiving programming over the wires were putatively in control of their identities). Receiving programming over the wires was so expensive that radio receiver manufacturers, who had had an early stake in making radio national, believed that they could not a fund national radio service. As broadcasting became increasingly networked, broadcasters looked to advertising as a source of revenue. This occurred in spite of the fact that in 1924, Secretary of Commerce Herbert Hoover had claimed, "if radio broadcasting shall be overwhelmed with advertising[,] the radio audience will disappear in disgust."[40] Throughout this era, the radio industry looked to precedents set by the railroads, telegraphs, telephones, electric lighting, and other utilities to keep its activities regulated, but not directly owned, by the federal government.

Even if one considers the technological choice that the industry settled upon to be a "technological imperative," many other factors influenced the shape of the national radio system.[41] For example, government approval sanctioned the wired wireless network system in the Radio Act of 1927. While not calling into question the format, ownership, or arrangement of the wireless network system, this act established the Federal Radio Commission and gave the government a one-year mandate to allocate frequencies. The 1934 Communications Act further entrenched and normalized this arrangement by not distinguishing networks or advertising-funded stations from other types of stations and not affording protection for religious, educational, farm, or labor stations.[42] The commission was "a weak agency, uncertain about its role."[43] This contributed to its readiness to reduce the complex and controversial regulatory issues it faced into "technical, instrumental problems."[44] Cultural historian Thomas Streeter writes that "within this world of experts, issues that to others might have seemed fundamentally political appeared as merely technical matters."[45] The FCC's readiness to constrain the spectrum allocation process to largely technocratic decision making resulted in the favoring of high-power national stations and gave small stations secondary status.[46] This set the form for what broadcasting remains today, even in spite of the expansion of FM and public broadcasting later in the twentieth century.[47]

Decades later, the activists and other advocates for LPFM (including some members of the FCC itself) asserted that the FCC's mandate included an obligation to promote localism as a component of its responsibility to ensure that broadcasters served the public interest. These critiques gained traction as the FCC deliberated over whether and how to establish the LPFM service. In comments filed in 1999, Prometheus argued,

Mandating Local Programming Is Necessary, and Does Not Constitute Regulation of Content.[48]

Stations should be locally programmed. No more than 20% of air time should come from off-site feeds or syndicated tapes. We feel that the bedrock value that the FM service was designed for, localism, has been undermined in recent years by changes in the broadcast industries.... We see LPFM as an opportunity for people to hear news and music and other programming from people in their own cities, towns and neighborhoods—an opportunity which is currently unavailable.

... It is not beyond the purview of the FCC authority to mandate local origination regulations, and they are clearly in the public interest.[49]

Other like-minded groups filed similar comments with the FCC during this period, claiming that LPFM was ideally suited to promote localism:

The nation's citizens have the right to exchange ideas and wrestle with local problems through the media. There is a compelling need for citizens to be able to get accurate, unbiased news and information about their localities, down to the smallest neighborhood. Low power radio presents this opportunity.

One significant reason to adopt a low power radio service is to improve coverage of local issues and culture. As the Commission recognizes, locally originated programming reflects the needs, interest, circumstances, and perspectives that are unique to the community being served by a licensee.[50]

Not only was LPFM held to promote localism, they argued that recent policies such as the Telecommunications Act of 1996 had harmed broadcasting:

Concern about profits in the commercial radio industry has all but eliminated local radio news reporting....

Localism will not be restored if low power radio becomes a replica of currently existing services. In this proceeding, the Commission has wisely decided to counteract the current consolidation in the full power commercial radio industry [permitted by the 1996 act]. Indeed, a primary benefit of creating a low power radio service is the likely increase in local coverage of local public affairs and culture....

While nationally distributed programming proliferates on radio and TV, locally oriented outlets are a dying breed. Increasingly, broadcast media outlets, including full power radio stations, are owned by large media conglomerates whose interests are far removed from the community a particular station serves.[51]

Here, these advocates argued that localism was part of the FCC's mission to protect the public interest and that radio was a medium well-suited to serving the needs of individual communities (though crucially, "localism" is not explicitly mentioned in the statute). They were met with mixed success. The FCC ultimately opted to consider locally originating content in the architecture of the LPFM service, but did not make it a strict requirement. Rather than mandating that locally originating content be required of all

LPFM stations, the FCC gave licensing preference to applicants who would broadcast a minimum of eight hours per day of locally originating content. This became an important consideration only when applicants for a given license were in competition with one another.[52]

Though the FCC was loath to dictate content, commissioners could make the case that localism was within the commission's mandate, because it related to diversity and the public interest. Former FCC commissioner Gloria Tristani (1997–2001, appointed by Democratic President William Clinton) served on the commission at the time of the creation of the LPFM service and commented in an interview on her obligation to the public interest:

> In my view the statute pointed to a balance ... you have to serve the public interest if you're using the public airwaves.... [T]he general powers of the commission section 303 ... section G [says]: "Study new uses for radio, provide for experimental uses of frequencies and generally encourage larger and more effective use of radio in the public interest." Read the law![53]

But she also said that she was influenced by her own experience with broadcast media after consolidation had occurred, following the 1996 act:

> I saw this massive consolidation and I saw the diversity of radio really being hurt [after the act]. I remember during the period that I was [on the FCC I was] traveling and you could switch cities, and [it didn't matter where you were], you would get into canned national playlists. Broadcasting—radio and television—it's fundamentally grounded in localism, or it's supposed to be. I just saw this system that wasn't remaining true to its mission.

Of course, one rationale for the 1996 act was that "new media" had deeply altered the terrain on which broadcasters operated. It was claimed that consolidation was necessary to allow radio to remain competitive. She recalled, "So then [we] had [the radio] industry countering, 'oh why are you worried about it? [people] have the Internet ...' But a lot of the product or content in the Internet is the same canned product, so you see that you're not getting the diversity and the use of spectrum that's intended ..." Although she said that localism was important in broadcasting generally, she endorsed LPFM in particular because of the diversity and localism of the stations that were on the air. She felt that LPFM's significance could not be captured by citing the raw number of new LPFMs alone:

> The amount of stations [has been] limited by what Congress did. So I don't think you can measure [success of LPFM] by the number of stations. But if you're going to measure success based on the stations that are there, to the extent you have empowered local voices to serve their community and be heard, I think that's a good

thing. It's a pretty diverse group. I know that about half of them are religious, but even so you have a very diverse group, diverse in every way. Student groups, ethnic immigrant groups, you have a very wide [range of citizens represented]—it would be so much wider if there were more stations.[54]

Tristani roughly equated pluralism and localism. For her, these both stood in contrast to consolidated broadcasting.

An aide to Commissioner Jonathan Adelstein (2002–2009, appointed by Republican President George W. Bush), stated very plainly that "locally originating content" was what made LPFM special and distinguished it as something worth protecting in the eyes of the commission:

What should be the benefits of LPFM is locally originated content, there is no question about it.... I think that's going to be truly a measure of their success and their long-term viability as well, both from a regulatory standpoint, meaning the FCC is going to continue to recognize and that there is this increased importance since there is this concern that a lot of that national, commercial full-power broadcasters are not providing the type of local content that's necessary.[55]

He reiterated the commission's disinterest in mandating content:

This is not content-specific at all, it could be specific religious programming that is not being transported from some other side of the country, or from some outside market, but is instead locally originated, that would be a true measure of success.[56]

It is worth noting that Tristani and Adelstein's aide specifically mentioned religious programming as a form of content that could "diversify" the airwaves. This indicated a reflexive need to paint LPFM as bipartisan and nonthreatening, to support free speech while steadfastly shying away from more "political" valences of broadcasting. The commissioners' nods to religious broadcasters may be explained in part by the influence of evangelical Christians in US politics during this moment. Regardless, touting "localism" was a way to appeal to groups espousing varying commitments and politics, without needing to reconcile otherwise divergent agendas.

Chairman William Kennard (1997–2001, appointed by Democratic President Clinton), who presided over the commission at the introduction of LFPM, also indicated that local use of radio was an important premise of the LPFM service. In a press release in April 2000, he stated, "While the National Association of Broadcasters frequently opposes new competitive services, I'm particularly disappointed that National Public Radio [public broadcasters] joined with commercial interests to stifle greater diversity of voices on the airwaves. I can only wonder how an organization that excels in national programming could fear competition from local programming by these tiny stations operated by churches, schools, community groups and public safety agencies."[57] Here, Kennard shamed NPR for siding with the NAB in the attempt to limit LPFM. He drew a contrast between the "tiny" LPFM's mission (which he characterized as harmless) and that of the national, networked broadcasters (both public and commercial).

Lawmakers also held that the promotion of localism would benefit the communities and constituents they served. The Local Community Radio Act of 2005, which sought to restore LPFM to the FCC's original intent, explicitly stated this:

(2) At a hearing before the Senate Committee on Commerce, Science, and Transportation, on June 4, 2003, all 5 members of the Federal Communications Commission testified that there has been, in at least some local radio markets, too much consolidation.
(3) A commitment to localism—local operations, local research, local management, locally originated programming, local artists, and local news and events—would bolster radio listening.[58]

The lawmakers who supported the expansion of LPFM argued that consolidation of the radio industry hurt localism in broadcasting and that the introduction of LPFM was a way to counterbalance that. In an interview, a staffer for Senator McCain (R-AZ) said,

The idea of another media outlet for different voices appealed to [the senator]…. The LPFM stations, some of them are being run by church groups where they're broadcasting church services for those who are home-bound, some are used for giving the most localized content and community information, some are used for reading storybooks to children to promote children's reading, some are used as a communications outlet for community groups that are trying to combat crime.[59]

The staffer indicated that a main attribute of LPFM was that it put a variety of ordinary citizens and community groups on the air. Mirroring the FCC, this staffer claimed not to favor any particular form of content. She instead viewed LPFM stations as potential resources for local communities, which could themselves determine the most appropriate use of their stations.

Station holders also offered examples of how stations were being used. They too cited the value of local programming to their communities. The founder of the Nashville station said, "The basic idea of [this station] was simple…. The airwaves belong to the people, and the people should have access to them. Everyday people have as much right to speak on the radio as to listen. I've found that Middle Tennesseans truly want their voices to be heard."[60] A little over a year after the Nashville LPFM went on the air, a volunteer commented on the station's commitment to local programming. She said that the station had a lot of "liberal" programming. But to her, the more important distinction was not "liberal" versus mainstream or conservative, but local versus nonlocal programming:

Nashville needs this. Not even a liberal outlet—we just need a community outlet.... There's no good news coverage. For example, there was a sit-in at the State Capitol. Tennessee was the first state to abolish state health care, so many of the people at the sit-in were really ill people, people in wheelchairs.... This event wasn't covered at all [by other news outlets]. The "local" NBC news affiliates [just] talk about crime and sensationalism. I hate the way liberal is coupled with good reporting. We just need local reporting.[61]

LPFM advocates strategically focused on the importance of local reporting in disaster situations. After the attacks on the World Trade Center and the Pentagon in September 2001, Prometheus organizers routinely stated that "when the planes hit the Pentagon on 9/11, the Clear Channel–owned station had to stream a CNN feed—there were no local broadcasters to report or give emergency information."[62] Prometheus seized on the potential utility of local media in terms of national security and preparedness. This claim was especially important as they made their pitch to lawmakers. After Hurricane Katrina devastated New Orleans in 2005, LPFM advocates intensified their call for the expansion of the LPFM service. They even obtained an emergency license from the FCC to set up a low power radio station (KAMP radio, 95.3 FM) in a trailer outside the Houston Astrodome, where hurricane evacuees were congregated.[63] Unlicensed microbroadcasters also went on the air to serve New Orleans.[64] In this e-mail, a Prometheus organizer solicited from an LPFM mailing list attestations about radio stations' activities:

Many of our stations did great work with and for the displaced families of the Gulf, after Hurricanes Katrina and Rita. We have a chance to tell the FCC how our community radio stations helped during the storms, helped rebuild after the crisis, and continue to provide donations, access and other support to groups even now. If we can make an impact, it could make a really big difference as the FCC considers important changes that can improve and protect the Low Power FM service!

... If your station (regardless of whether or not you are located in the Gulf region) did anything to aid in the relief efforts for hurricane Katina, please write a summary of what you did and send it to Prometheus so that we may bring these stories to the attention to the FCC.[65]

A Prometheus press release discussed the legislative efforts to protect and expand LPFM on the basis of its use to disaster response teams and the public during emergencies. Specifically, Prometheus told the House of Representatives' Subcommittee on Telecommunications that language to expand LPFM should be added to the 2006 House of Representatives Warning, Alerts and Response Network (WARN) act:

Across the Gulf Coast and in countless other situations across the country, locally owned, volunteer-run community radio stations like LPFMs have been the difference between life and death, safety and danger for local communities.... Stations like WQRZ-LP, a low power station in Bay St. Louis, Mississippi, were perfectly placed to give neighborhood-by-neighborhood coverage of the damage of Hurricanes Katrina and Rita, and to help those communities know exactly how to interface with local and federal safety and health officials after the storms.

But these stations are few and far between—limited from thousands more towns and neighborhoods by an out-of-date law limiting low power radio to small, remote communities.... If Congress moved to expand low power FM radio to thousands more towns and cities across this country, these communities would enjoy a reliable and well-understood local technology when disaster struck.[66]

The LPFM advocates argued that consolidated broadcasters were not well-suited to provide disaster coverage or local, up-to-date information. This was because they might not have staff who knew the area. They might not even have human staff members on site at all, because much programming is automated. Instead, LPFM advocates recommended that safety and law enforcement officials forge relationships with LPFMs. In many cases, they could guarantee that live people with intimate knowledge of the area would be accessible at the station in times of emergency. The LPFM advocates' positioning around the subject of disaster preparedness distinctly echoes arguments made by ham radio operators in the 1940s. Hams offered to conduct public alerts or transmit vital information for "citizens' defense," in exchange for being allowed to stay on the airwaves during World War II.[67]

During a workshop on emergency preparedness at a barnraising, one LPFM station operator reflected, "It scares the heck out of me that we could have a disaster in our town. I want to know exactly what to do in case of a flood, earthquake, or whatever. We are the station they [the listening public] would turn to."[68] Another workshop participant said, "If you're the station that's on the air with good information, you'll increase your listenership and loyalty to the station. As opposed to Clear Channel or whatever, with their automated programming."[69] This volunteer also recommended that LPFM broadcasters pursue ham operator training in order to feel confident technically as well as to have access to the ham network in case of an emergency. Privately, Prometheus also advocated that LPFM operators buy transmitters that could operate above 100 watts "just in case." One volunteer at the emergency preparedness workshop said darkly, "Shortwave amateur is the last backup [for a communications system] if everything else fails, and the way this country is currently moving ... it's not outside reality [that a backup communications system might be needed] in my way of thinking."[70]

LPFM advocates also underlined localism in musical programming in the wake of commercial radio's consolidation. According to these advocates, consolidation affected musical artists economically and reduced the public's access to musical diversity. The Future of Music Coalition, a nonprofit think tank and advocacy group focused on independent music and technology, stated in a press release that "radio ownership consolidation at the national and local levels has led to fewer choices in radio programming and harmed the listening public and those working in the music and media industries, including DJs, programmers and musicians." The press release went on to add that "commercial radio now offers musicians fewer opportunities to get airtime and offers the public a narrow set of overlapping and homogenized programming formats."[71] The volunteer from the Nashville LPFM raised issue of cultural programming, in addition to news:

> Nashville is such an arts and music town. People are so hungry for [a media outlet] that they'll travel for 45 minutes [to the station] just to broadcast their music.... The station has radio plays, a theater group does a radio play twice a month. We have lots of funny, weird programming that makes community radio and public access so good. The music—we have old country and bluegrass, that's cool because that's really what Nashville *is*; the commercial country stations are just commercial new country. We have hip-hop shows, French and Cambodian music shows, a 70s funk show, an indie/punk/hardcore show with obscure awesome records, Cajun music.[72]

For this volunteer, the eclecticism of the music offered on the station was a benefit. She contrasted this programming to commercial radio, stating that "funny, weird programming" is largely limited to community media. Many people at the Nashville barnraising mentioned that the city did not have any other outlets for local styles of music such as old country and bluegrass. Prometheus referred to a similar situation in Opelousas, Louisiana, where they held an LPFM barnraising in 2003: "Opelousas is home to Zydeco—hard-rocking accordion, fiddle, and guitar music native to Creole country like Opelousas. But because of the dominant presence of consolidated commercial radio in Opelousas, Zydeco was nowhere to be found on the radio dial—until KOCZ came along."[73]

Some religious advocates for LPFM also drew attention to localism. Prometheus somewhat uneasily allied itself with some Christian broadcasters in hopes that they could pressure their elected officials to side with LPFM and against the broadcast lobby. Prometheus knew they would not sway politicians in conservative areas. But they hoped to leverage the relationships religious (particularly evangelical) constituents already had with elected officials in conservative districts in order to gain traction for LPFM. They also wanted to make LPFM appear as bipartisan as possible, with strong support on both sides of the aisle in Congress.[74]

Syndicated religious programming is a large, well-established, and lucrative phenomenon. The broadcast lobby represents many religious broadcasters as well as secular ones.[75] Prometheus hoped to strategically spin the LPFM issue in conservative districts by convincing elected officials to serve their own local constituencies as opposed to siding with syndicated religious broadcasters and the NAB. Some Christians agreed: "It looks like money's talking, maybe at the expense of what would be logical [in the] long term—the Republicans supporting their supporters," said a spokesman for the National Association of Evangelicals.[76] An advocate for Christian LPFMs and community broadcasting said in an interview, "whether it's a small church or a small advocacy group, it's impacted by big full power broadcasters.... It's not an issue of all blue people [Democrats] want one thing and all red people [Republicans] want another. Whether control is by monolithic Christian entity or monolithic [Clear Channel], those are antithetical to the interests of a local, small-scale religious broadcaster. It's not a theological or a classic 'whether the programming on the station is going to be liberal or conservative' issue."[77] He stated that LPFM was ideally suited to small churches that could run their own original content twenty-four hours per day, using an all-volunteer staff, for less than $1,000 a year. He gave some examples of the types of programming employed by these stations:

> [There is a] program in Florida, secular music except one minute an hour they talk about an issue to think about.... [I know a] church station that does funeral announcements, and they sometimes get a call complaining "why didn't you do announcements?" The answer is, No one died! The church groups that I know do not do it to run announcements over and over saying "come to our church" [proselytizing]. They want to reach with positive programming a target audience in the homes.[78]

This advocate held that the main issue for religious broadcasters and for secular community broadcasters was local programming that small groups could choose for themselves and actively participate in creating. To him, this local emphasis was more important than the theological or political basis of the content to be broadcasted. His anecdote about the funeral announcements highlighted the autonomous, highly local, even idiosyncratic programming LPFM advocates often favored.

Commenting on Prometheus's relationship with proponents of Christian community broadcasting, one Prometheus organizer said, "Prometheus cleans up nice, but we're not politically able to get senators like Sam Brownback [R-KS] or Trent Lott [R-MS] or even Kay Bailey Hutchison [R-TX] to talk to us as openly or as communally as we'd like.... In order to be able to expand low-power FM on the Hill, we're going to need these people."[79] In an e-mail, Ellen described her experience meeting with a group of Christian allies: "the christians were okay—they actually tried to convert me, which is insane. i made good contacts with broadcasters who have actual relationships with their elected representatives, which is the point ..."[80] She was mildly uncomfortable, but political expedience won the day.

In a meeting with progressive lawmakers, Jasper stated that LPFM "is an important opportunity to take representatives and senators to task for siding with corporations and against constituents like small local conservative churches."[81] But Prometheus's main interest was not in promoting the rights or agendas of conservative religious broadcasters. For Jasper, this was a means to an end. He hoped that Prometheus would succeed in their goal of establishing "progressive radio not just in the big cities but into every nook and cranny around the country" (as noted previously). (Progressive LPFM might well include progressive Christian groups, but a major divide the activists hoped to span in Washington was between conservative, religious advocates for LPFM and secular liberal ones.) Interestingly, the aforementioned advocate for Christian community broadcasting speculated that a motivating factor for the passage of the 2000 rider limiting LPFM was Republican paranoia over giving broadcasting licenses to groups in inner cities: "[The] reason the December 2000 rider lines passed, to limit LPFM, what is the underlying non-technical reason? This [was] a Republican Congress. If you limit stations, fewer stations in big cities, cities are traditionally more Democratic, radical left-wing, [the Republicans didn't] want to give licenses to the groups in inner cities, since they don't believe that inner-city churches are the ones who are going to get the licenses."[82] He was resentful that Republican distrust of urban Democratic populations had an impact on the ability of urban church groups to procure LPFM licenses.

National Public Radio, however, was more ambivalent—some might say cynical—about the topic of localism. NPR acted to oppose LPFM, siding with the NAB on the issues of potential interference and the need for third-adjacent channel protection. NPR also supported the Radio Broadcasting Preservation Act of 2000. At the same time, NPR argued that in principle, it was in favor of the addition of more community radio stations. NPR president Kevin Klose stated, "Some in the industry wanted to gore low-power radio, we didn't do that. We aren't criticizing anybody. We have a set of very reasonable issues [about interference] that can be addressed. The goal is a conflict-free future between low-power and public radio."[83] NPR's claim to seek a "conflict-free future" with LPFM seemed rather disingenuous given that elsewhere, it argued against the introduction of LPFM, claiming that full-power radio stations and the Internet could better serve the public than LPFM. NPR stated that "it makes no practical difference to the listener whether the source of the content is a low-power station transmitting from a mile away or a full service station transmitting from five or ten miles away."[84] Through this statement, NPR sought to deflate a central claim by LPFM proponents that low-power and full-power stations would not serve the same functions. NPR, in essence, claimed that LPFM provided nothing NPR did not already do and that NPR did it better. (It is worth noting that NPR glossed over whether "local content" meant content that served a local community's needs or content created by a local community and originating at the site of transmission.[85])

While they hailed "localism" and attached this value to LPFM, many advocates for LPFM disagreed on the particulars of what was meant by this. Various interpretive communities with differing political commitments each mobilized localism in their advocacy for the creation and expansion of low-power radio. Differing rationales for the importance of localism included a general public good arising from coverage of local news and affairs, the promotion and sustainability of musical artists, public safety and disaster preparedness, and the value of community autonomy. There was also a "for-granted" notion of localism that was not otherwise specified. In all cases, local programming was contrasted with syndicated programming carried by either commercial or noncommercial stations. I do not wish to elide very real variances in interpretations of localism or the reasons for deploying the concept differently in various settings. Rather, my goal in this section has been to show that whatever their differences, many groups mobilized localism as an interpretation of LPFM.[86] Perhaps needless to say, the value of "localism" was inscribed in the 2011 act that expanded LPFM, because it was titled the Local Community Radio Act.

Noncommercial Status

At the inception of broadcasting, it was not clear how radio would be funded. Broadcast historians have demonstrated that the dominance of commercial radio was not inevitable. Every step involved conflict.[87] The pattern that stabilized over the first fifteen years or so of broadcasting came to favor a national network system that paid for itself (and indeed turned a profit) through advertising. National networks also created economic efficiencies around distribution of programming. "[T]he resulting flow of money swept up nearly everything in its path, so the majority of stations became commercial operations within a few years and network affiliates soon after,"[88] write media scholars Eric Rothenbuler and Tom McCourt. Commercial broadcasting was durable, remaining ascendant even decades later.

In the first decade of the twenty-first century, as the radio activists labored to expand LPFM, broadcast media still commanded the nation's attention. Statistics from 2004 indicate that radio was Americans' most popular form of media after television, with people spending an average of 983 hours listening to the radio annually (compared with 1,661 hours of television).[89] According to communications policy scholar and advocate Mark Lloyd, the advertising revenue for radio was $17.86 billion in 2001. This figure was slightly higher than the amount of advertising revenue earned by syndicated network television programming. (This number excludes station affiliates and cable television, which together earned another $37 billion.[90]) Though television had indeed eaten into radio's primary position by the 1950s,[91] the clout of radio remained significant in terms of listenership and advertising revenues. Radio advertising revenues fluctuated over the first decade of the twenty-first century, falling some but then regaining traction during the economic recession at the end of that period. Radio's audience remained substantial, averaging around 233 million listeners per week in 2007–2008.[92]

Out of twelve thousand full-power stations nationwide, in 2005 there were around two hundred full-power community radio stations (noncommercial, non-networked) and another eight hundred National Public Radio stations (noncommercial, independently operated, but affiliated with NPR).[93] Commercial full-power radio stations outnumbered the noncommercial ones twelve times over. The FCC and other groups have regularly noted that noncommercial radio is an alternative form of radio with great potential to serve the public interest. The radio activists strongly supported the notion that radio should be not for profit. They felt that a radio station without a profit motive was inherently different from a commercial station.

In the era between the raid of the Radio Mutiny station in Philadelphia and the initiation of LPFM, Jasper submitted the following comments to the FCC. In no uncertain terms, he took aim at incumbent for-profit broadcasters:

> I would also like to take the opportunity to point out that the naked self-interest invoked in the comments of many of the [incumbent] broadcasters.... The Telecommunications Act [of 1996], and the public pronouncements of all the Commissioners, have mandated the promotion of competition. Yet the comments of the licensed broadcasters complain bitterly that microradio could cut into their marketshare [*sic*]. Give me a break! You've spoiled these broadcasters rotten! They are so accustomed to the conditions of regulated oligopoly that they are afraid of competition from

10-watt community radio stations. This obviously has nothing to do with orderly regulation, interference or spectrum scarcity—this is regulation to protect the rich from the poor.[94]

He also chastised the FCC for exalting radio's potential to make money at the expense of other interpretations of broadcasting (such as micro-powered community stations). Notably, Jasper's comments illustrate not only his sympathy and identification with pirate broadcasters but also actual pirates:

I take the opportunity of these comments to bring you the words of Captain Avery, a renowned pirate.[95]

"damn ye, you are a sneaking Puppy, and so are all those who will submit to be governed by Laws which rich Men have made for their own Security ... but damn ye altogether: Damn them for a pack of crafty rascals, and you who serve them, for a parcel of hen-hearted Numskuls [*sic*]. They vilify us, the Scoundrels do, when there is only this Difference: they rob the Poor under the Cover of Law, forsooth, and we plunder the Rich under the Protection of our own Courage."

Commissioners of the FCC, I ask you this: will you go down in history as "a parcel of hen-hearted Numskuls?" Will you use this rulemaking [to create LPFM] as a cover for your campaign of repression against us, or will you make a real change? ... Unlike Captain Avery, we are not even "plundering the rich"—the citizens of this land do not owe the commercial broadcasters and advertisers our listening attention. Services of the nature of microradio were legal in this country before and have some analogues in other industrialized nations. It's time for some balance, and an end to the regulator's slavish indulgence of the broadcasters [*sic*] oligopoly.

In his invocation of the broadcasters' "oligopoly" and the FCC's "slavish indulgence" of it, Jasper made it plain that for-profit broadcasting was not his favored use of radio. But he did not challenge the right of incumbent broadcasters to operate on a for-profit basis. Instead, he cited the worth of broadcasting for purposes other than profit accumulation. (Notably, non-commercial full-power licenses are usually traded at auction, keeping them well out of reach for most individuals or community groups, in contrast with LPFM.) Other comments to the FCC echoed some of these sentiments; the United Church of Christ wrote, "low power radio will help to balance the service provided by large commercial radio broadcasters. As such, a noncommercial service will be the best counterweight to a commercial service that chooses to serve audiences based on their ability and willingness to buy products. Furthermore, many of the entities that are well-positioned to begin a small low power radio service ... are non-profit organizations."[96]

LPFM's backers objected to the relationship between media content and the commercial interests of for-profit media (or parent) companies. Jasper stated, "Many of us in the microradio movement see the current round of consolidation of ownership in media as potentially doing great harm to what is left of the democratic institutions of our society." Specifically taking up the issue of corporate ownership, he wrote,

A serious issue is the interlocking financial relationships of news media outlets and organizations with other financial interests. It would be foolish to try to get unbiased reporting of issues around the nuclear industry from CBS, which is owned by Westinghouse, which is a leading manufacturer of nuclear power plants. Of course, we do have many choices in America's media today: for example, you could look for your information about nuclear energy from another network, such as NBC—the only problem is that they are owned by General Electric, another major manufacturer of nuclear power plants. Media must be independent to be credible.[97]

Jasper opined that for-profit media companies simply could not fulfill their obligation to robustly serve the public interest in their news and public affairs reporting. To do so would create a conflict of interest for them. Instead, he proposed that an alternate form of media—independent, noncommercial—had a different and valuable role to play. Without a profit motive, and without being beholden to the political or financial interests of a parent company, noncommercial media would have a better opportunity to conduct news investigations and reporting free of bias.

LPFM supporters also contended that musical and cultural programming were qualitatively different and better on stations without a commercial motive. A Future of Music Coalition advocate stated,

Part [of our work on radio] is a constant critique of the commercial radio system, of consolidation, issues of institutional Payola[98] ... A big part of it for us is understanding and appreciating the value of noncommercial radio ... Noncommercial radio fundamentally is saying, I as a disc jockey, I as a program director ... I want to bring something to my audience. I want to play this song because I love this song, I want people to feel about this song the way I do ... That's the fundamental distinction, commercial radio is a business, [while] noncommercial radio at a very basic level is spreading culture, it's using radio as an art form to disseminate a piece of culture.[99]

He was quite specific about the ways in which he felt noncommercial radio constituted a valuable resource for musical artists and listeners:

Noncommercial radio is where you see less of the regimented formatting, less of the specific carving up of demographics into age and race and gender, a greater appreciation and openness to other formats, opportunities for local or regional musicians, a home for independent music.

Music released on independent labels makes up 80% of the music sold in this country, but it's virtually absent from commercial radio ... if you're [recording] on an independent label you can go to noncommercial radio. LPFM obviously fits into that context.... There may not be a business case to run a 50,000 watt station that's gonna play Pavement and Yo La Tengo [popular "indie" bands]; however there are college stations all around the country that play [them]. And of course I'm dating myself! We could update that to say Wolf Mother ... or *Destroyer's Rubies*, Dan Bejar made the great record of the year and it's selling very well and it's on Merge, which is a big record label, but you don't hear it on [commercial] radio....[100]

Our feeling with LPFM has always been that all we've wanted to do is create the platform.... We're not saying that we want LPFMs set aside to make up for the inadequacies of our radio marketplace as it relates to independent musicians, we just want to create the platform and let the great diversity of the American public come through in that.[101]

According to this advocate, noncommercial radio stations could make freer decisions about the music they played because they were not restricted to playing music selected for its appeal to a particular demographic. (Local and regional music, which he also cited, may be similarly narrow in its appeal.) This enabled stations to play music for no reason other than the DJ's individual taste or curiosity about unknown music. This had the added benefits of challenging listeners with novel musical options and diversifying the cultural landscape.[102]

Many argued that noncommercial radio provides economic and cultural benefits to society, exposing the public to a range of musical artists they would not encounter on commercial radio. They held that such exposure was essential for a thriving musical landscape in which musicians could earn a living making music. Prometheus sought academic ballast to lend credibility to assertions they made about the impact of low power and noncommercial radio on musicians. They distributed a handout soliciting scholarly help to study these issues: "Does the station play or promote local musicians who might otherwise not have an outlet? Does this affect sales, or local knowledge of and/or participation in local music events?"[103] Speaking generally, scholars have often argued that "diversity is a social good valued in its own right ... in the case of organizations whose outputs are largely symbolic or cultural."[104]

Radio's ascent as a medium that provided specialized formats was closely tied to the rise of market research in the post–WWII period. Along with television, radio became a "demographic vista."[105] As revealed in the advocate's comment about "carving up the audience into demographics," many LPFM proponents were skeptical of what they saw as the reduction of radio broadcasting to market-based metrics and values. They felt this limited the conversation about radio. When I began fieldwork in 2003, I was initially puzzled when some activists suggested in interviews that the notion of who was listening was something they rarely considered. Although some people described how gratifying it was to receive a phone call from a listener with feedback about the programming,[106] others stated the act of transmission itself was more significant: "I think that there's something really powerful for the people that participate [in broadcasting], it's empowering, people learn a lot, it builds community, there is a social value just to broadcasting, I've seen people get changed by the experience of being DJs, being journalists, and that's really positive."[107] Others scoffed at radio being "ruined" by endless attempts to scientifically cater to varying demographics.[108] In this stance, activists critiqued commercial broadcasters' practice of carefully measuring listenerships, which is necessary to generate advertising revenue.[109] They were leery of the profit motive generally and of the notion that these sorts of metrics were of much value in understanding the true potential of radio.

Even in the realm of noncommercial radio, the activists were dubious about market economy metrics as a primary means of considering the use and impact of radio broadcasting. The CPB (a source of public funding for noncommercial stations including NPR) gauges the merits its applicants by evaluating fundraising and listenership. Particularly as LPFM aged, activists were keen to generate data that would justify its continued support by regulators and funders (as well as its expansion by lawmakers). At the same time, activists were in a bind because they felt that the CPB's metrics were unable to fully capture and express the value of LPFM stations. They contended that market-based metrics did not indicate the value of a station to its listening community. This was especially the case in poorer communities where donations to the stations were lower.[110] In 2004 comments to the CPB, Prometheus wrote,

> What is missing in the CPB funding scheme is funding for small, community based, projects driven by volunteers.... [Does National Public Radio] at this juncture in history need every available dollar, while emerging LPFMs are ineligible for even a nickel of CPB funding?[111]

The anger at NPR is palpable here. The activists were correct that the scale of many LPFM stations made them all but incomparable to larger-scale NPR affiliate stations.

Elsewhere, the radio activists again raised the issue of audience size being an inadequate measure of the impact of LPFMs. They argued that valuation of audience size was tied to commercialism, and thus it undercut a more robust consideration of public interest values:

The Corporation for Public Broadcasting currently uses two measures of "audience service" in its Community Service Grants. One is the measure of audience numbers as implemented by Arbitron [a listener data market research company].... [A] problem is that while audience numbers are measured by Arbitron, impacts are not. It could easily be argued that there is more benefit to society from certain types of programming and station organization than others. Even if Howard Stern telling fart jokes attracts ten times as much audience as a broadcast of the city council meeting, the positive impact on society may be greater from the availability over the airwaves of the deliberations of government.

Additionally, these numbers are collected in order to be used by the advertising industry, which means that their research design may or may not be appropriate for community radio use.[112]

Plainly, the radio activists felt that the metrics used to evaluate commercial radio and full-power noncommercial stations such as NPR affiliates were completely inappropriate for LPFM. In particular, they held that these measurements ignored the special properties of small-scale, noncommercial stations such as LPFMs and some community stations: "Measures [that] have sometimes been adopted for evaluation ... said more about what was convenient to measure than about the actual impact of community radio."[113]

Programmers of LPFMs also expressed a schism between professional, for-profit radio values and those of some LPFMs. A Prometheus volunteer said in an interview,

at its best, community radio manages to avoid a lot of the nasty side effects of commodification of media—most media products are designed to be a commercial product, to give you what you expect and what is comfortable, to be effectively associated with whatever products are being advertised, [which can] have the effect of dumbing things down, making things bland, formulaic ... [whereas] community radio can have a new breath of originality, can be locally relevant, can reflect the values of a community, can be interesting.[114]

Maria, a volunteer at the Coalition of Immokalee Workers' LPFM station in Immokalee, Florida, compared her station to commercial ones in her area. She claimed that the adjacent commercial stations perceived that her LPFM had taken away some of their listeners, so "they made changes and started sounding more like us, like playing Guatemalan marimba music. But we don't think of ourselves as competition, we're independent, we can [still] do things they can't, like we'll talk about the same [issue] for two or three hours or two or three weeks if we want to." She also said that "community radio is not professional; it feels more comfortable."[115] The program director for the Oregon farmworkers' station said that he would specifically be interested in "not professional" volunteers such as youth in schools becoming programmers.[116] This should not be taken as an indication that all LPFMs are anti-professional in their programming or production values (though there is often an internal debate at community stations about how "professionalized" they want their programming to feel and sound). In fact, LPFMs and community stations argue that they provide training grounds for people who will later find employment in the broadcast industry. Professionalism is by no means demonized. But the statement of the programmer from the Florida station is telling: the station would cover an issue as long as they felt it was necessary or relevant, and they did not feel the need to justify that decision commercially.

In response, the NAB argued that commercial stations were better able to serve the public by operating on a for-profit basis. At the heart of many of the NAB's statements to the FCC was the belief that profitability in broadcasting was an asset that served the public good. NAB also constructed profit as an inalienable right of broadcasters: "The Commission has recognized that 'the industry's ability to function "in the public interest, convenience, and necessity" is fundamentally premised on its economic viability.'"[117] The NAB insisted that signal interference would result in loss of listenership for incumbent broadcasters. This would result in revenue loss. They further claimed that even if signal interference did not occur, LPFMs could create a negative economic impact on incumbent stations due to "audience diversion" (which they insisted would particularly affect small and minority broadcasters).[118] They also held that their member stations' for-profit operations enabled them to provide better programming to increasingly diverse market segments. In an interview, an NAB spokesman stated,

> There are critics who say that of the 1996 act [allowing consolidation] resulted in bland and boring programming. I simply disagree on that, I think there's never been more diversity on radio than there is today.... 60 percent of all radio stations were losing money in the early 1990s. You can't provide a good service to your listeners if your station is going bankrupt. Congress understood that, they recognized there was a reason to allow some consolidation.[119]

In the NAB's assertions that LPFM would hurt incumbent stations' economic interests, the implicit model of radio was that radio was a tool to make money and consolidate audiences. (If history is any indication, NPR may agree with the point about audiences, if not the profit motivation.)

Conclusion: A Toaster with Sound?

As we consider the empirical examples of how various groups including the activists attempted to shape LPFM, meaningful contours emerge. At the most general level, we hear how broadcasting was open to reinterpretation decades after its dominant meaning had stabilized. Old technology does not cease to develop just because it is old. In fact, the interpretation of technology is subject to ongoing construction, maintenance, and repair—a point we miss if we train our ears only on new technologies. With a technology as rich and old as radio, a complicated range of meanings may form. This dynamism over time has implications for both scholars and propagators of technology.

The activists differed from other groups in terms of where their emphasis lay. The three main interpretations they championed—political organizing, localism, and noncommercialism—represented a challenge to longitudinally dominant understandings of radio. The activists had to push for noncommercialism and localism in LPFM. In shaping the new service, the inscription of these values into LPFM was not taken for granted. The interpretation of radio as a tool for political organizing was one that was more contentious. The activists often kept this idea to themselves when making policy arguments (though in certain circumstances, particularly with progressive lawmakers and community groups who wanted radio stations, this use of radio was highlighted). Notably, the FCC shied away from advocating for the political use of the stations and refused to privilege some groups' content over that of others when allocating licenses. By contrast, localism was a curiously noncontentious value, even when groups plainly did not agree over what was meant by "localism."

This work to construct and contest interpretations of radio also highlights the activists' roles as mediators of technology. In this chapter's empirical examples, I show how the activists attempted to speak for the interests of the public and community groups. I also illustrate the activists' attempts to shape interpretations of radio held by crafters of policy, producers of radio, and consumers of radio; the activists argued for and against certain interpretations of radio, and attempted to persuade others.

Media activists argued that ordinary people ought to "become the media."[120] They proclaimed that "media should be a tool for communication and transformation and not a commodity to passively consume."[121] This sentiment has precedent; contestation over the reduction of listeners to "consumers" dates to the earliest days of the medium.[122] The radio activists' attachment to the notion that radio should be noncommercial should be taken as a profound critique (in line with their broader activist ideals) of the notion that media should be geared toward a buying public. This critique, especially when combined with the values of localism and political organizing, had deep ramifications. It was not merely a strategy for expanding the use of media. It was tied to beliefs about promoting community self-determination and mounting a challenge to neoliberal globalization. These activists believed an independent media system was of paramount importance in that struggle, and the goal of social justice was indeed what animated much of the movement for media democracy.[123] In addition to opposing consumerist values,[124] the radio activists also resisted a technocratic discourse in their interpretations of radio. In the early history of FM, regulators opted to focus on "technical, instrumental problems" in decision making.[125] In so doing, they sidelined complex social and cultural issues. In promoting localism, noncommercialism, and political use, Prometheus activists worked to insert nontechnocratic values into decision making about FM radio. In the process, they challenged regulators and incumbent broadcasters. Debates that appear to be "settled" in the early days of a technology may not stay stay settled, which a technology as old and freighted with meaning as radio makes abundantly clear.

Media activism can contest the structure and social relations of media systems, the codes of communication produced within media systems, or both.[126] However, contradictions emerge between media structure and codes. On the one hand, the radio activists seemed to regard the propagation of LPFM to be inherently liberatory. This was evident in their willingness to assist nearly any group who asked for help preparing a license application or building a station. In many situations, they advocated for universal availability of LPFM. This was how they approached lawmakers in particular as it enabled them to present LPFM as bipartisan. The activists made the case that small-scale radio was suited to use by community-scale groups, whatever type of organization they might be. On the other hand, in other instances they were candid about their desire to see LPFM stations in the hands of groups with politics they favored. They hoped to see these groups use LPFM stations for political organizing and for the creation of media to support social justice work. In the aggregate, it is difficult to ascertain precisely what the radio activists understood to be the consequences of their propagation (other than an often unspecified positive effect).

In their advocacy, the activists studiously ignored a paradox. In the main, they held that small-scale radio technology was not a "neutral tool." They believed it had inherent effects, predominantly the promotion of community autonomy and fostering of democratic social conditions. But they glossed over the possibility that widespread distribution of LPFM technology could ever have regressive applications (including use by groups with whom they disagreed politically). These topics are worthy of consideration. However, the issue is thornier than one of "positive" versus "negative" "effects." To the extent that consequences of LPFM might be identified, did they flow from the diffusion of the technology itself? Or from its use to promote certain messages? And what are the important analytical issues at stake in this distinction (if there is one)?

Media activists are not alone in running afoul of the issue of "content" versus "artifact" when considering the use and import of media technology.[127] FCC commissioner and Reagan appointee Mark Fowler, a deregulation enthusiast, sparked controversy with his 1987 statement that "a tv is a toaster with pictures."[128] Advocates for strong public interest policies countered that "media are not merely appliances to be used by consumers. Media constitute a central political and social institution, one that is essential for the construction of citizenship and the maintenance of a healthy public sphere."[129] Fifteen years later, Commissioner Jonathan Adelstein cleverly engaged Fowler's statement by extending his metaphor, stating that "if [TV] is the toaster with pictures, soon only Wonder Bread will pop out [if further consolidation is allowed]."[130]

Though Adelstein's remark pithily skewered proponents of consolidation, his challenge to the "toaster with pictures" does not resolve how analysts should treat media and information technologies. Activists, regulators, and other groups routinely draw on the idea of radio as a conduit for content. They construct a boundary between the content and artifact, or message and medium. But they are at other moments comfortable generalizing about effects flowing from radio itself. This contradiction is interesting because it shows the difficulties reflective and sophisticated actors such as regulators and radio propagators (as well as analysts) have in distinguishing "content" from "form." We might reframe this issue by considering "content" as simply a concept around which actors interpret and negotiate technologies.[131] This allows us to acknowledge that both "message" and "artifact" are regularly at play in our thinking about media technologies without committing us to a notion of inherence.

Overall, the radio activists aimed to change far more than the messages that were transmitted by radio stations. They sought to challenge "the entire complex of social relations and practices through which [messages] are produced and disseminated."[132] The activists' understanding of what radio was, and their feeling that its diffusion could foster democratic social relations, was consistent with their desire to propagate radio technology.[133] Positing an alternative future for radio, the Peruvian activist stated this forcefully: "We need to strive to turn community radios into actors—not only spectators, but actors in society; not only informing others what is happening, but transforming the reality.... Here we say and believe that a different world is possible. We also say and believe that a different radio is possible." Likewise, the Nashville station holder echoed the belief that broadcasting should include everyday people. She also supported active transmission as opposed to passive listening: "The airwaves belong to the people, and the people should have access to them. Everyday people have as much right to speak on the radio as to listen."

7 Do New Media Have Old Politics?

Reflecting on the earliest organizing he had done around LPFM, a Prometheus organizer cited Mbanna Kantako. Kantako was an early hero of the micro-broadcasting movement who had broadcasted without a license from a public housing project in Springfield, Illinois, in the 1980s. He devoted his broadcasts to social justice topics including police brutality, racism, and poverty. Kantako understood his transmissions as electronic civil disobedience and "a potent means of regaining power and a voice within an oppressive local system."[1] The activist said that this story was one that filled him with passion, and that although he wanted to capitalize on this story to provoke an enthusiastic reaction in others, he did not wish to exploit Kantako. He recalled, "I told his story over and over, like 200 times, in the most respectful way I knew how."[2] In the main, he felt he had struck the right chord: the early lawless origins of LPFM made it "a hell of a story, it captured people's imaginations."[3] In the mid-2000s, Prometheus was poised to expand from radio into community wi-fi. But this was not a seamless transition. The same activist said that his main problem with the group's work on wi-fi was that unlike radio, which he considered to be an issue people were willing "to go to jail over," he didn't know how Prometheus could convince anyone to "fall on a bayonet" for wireless.[4] He said, "we need to take [wireless] into people's hearts [and make them see that] it's not about getting a cheaper cable bill [from an Internet service provider] ... we need to seek danger."

This chapter follows the interplay between radio activists' assessments of FM radio and emerging Internet-based technologies, primarily wi-fi networks. It highlights their role as mediators of technology, engaged in a quest for "appropriate" technical options at the community level. They promoted these options to would-be users and to regulators as well as other groups with the power to influence policy or shape technology. How did the considered negotiation of new technology play out? What were the benefits to holding a focus on radio in order to understand Internet-based technologies? In practice, the activists largely rejected some proposed technical alternatives (such as webcasting), yet they cautiously embraced others (such as community wi-fi networks).

As the activists assessed Internet-based technologies, they worked to translate the perceived assets of LPFM into the domain of emerging technologies.[5] They sought to retain the vision, flavor, and organizing strategies from their LPFM campaigns while circumspectly negotiating the expansion of the organization's efforts to include community wi-fi networks. This process highlights a nuanced interplay between "new" and "old" media.[6] (We might miss this complexity if we listen only to new technology.) Radio is an entrenched and well-understood artifact. Such already-developed technologies provide insight into how new technologies are interpreted and taken up. This book holds that the significance of new and emerging communication technologies can be grasped most effectively when emerging technologies are considered in a dynamic field that includes older technologies.

The activists puzzled over how to gauge which technological options were best suited to particular purposes and how to shape interpretations of technologies. They also confronted their own role as mediators. Differences in position and privilege meant that some would-be users did not understand technologies in the same terms that the activists did. When this happened, the activists' technological mediation was further complicated. Attempts to incorporate other technologies once again led the activists into collision with historical patterns of inclusion and exclusion, including race and paternalism.

Activists aimed to assess changing technological options in light of both present and future needs. They also exhibited a strong awareness of the past. One policy advocate said, "I think we are at a good point in telecommunications policy and technology.... It hasn't been this way since the 1920s[; now] we have an opportunity to secure spectrum for people beyond businesses. The window will close again within two to three years and be closed for at least another 70 years."[7] Prometheus organizer Ellen claimed, "There is a sense of urgency because of the new technologies. The Telecommunications Act is being rewritten."[8] Scholars have argued that the policies implemented in the 1920s and 1930s profoundly affected the media landscape for many decades.[9] The activists were largely conversant with these arguments, and cultivated a deliberate historical awareness. They believed that their technical and political choices about media technologies were important because of their implications going forward in time. As one activist stated, "The idea of spectrum scarcity is changing fast, and it's up to us to understand technologies.... The ideological struggle is whether Verizon will own the spectrum and sell it to people, or whether the spectrum will be unlicensed and available."[10]

The activists realized that the demand for LPFM radio stations (or even other terrestrial radio[11]) was not endless. Technical and political changes of various sorts would inevitably reshape the media landscape. Prometheus believed that the ability to add more FM stations to the dial was limited. One of Prometheus's board members said in a 2005 meeting, "LPFM is finite, in five years all the stations [the FCC will license] will already be on the air or won't be able to get on the air."[12] With the passage of the Local Community Radio Act of 2010, LPFM was further expanded. But the general principle remained that access[13] to the FM spectrum was circumscribed. As a consequence, the ongoing feasibility of their focus on FM was brought into question. In thinking about the future, one activist indicated that FM radio would possibly be less relevant to the organization:

> Prometheus is working for social movements we believe in and to democratize technologies. Wherever there's a communications technology that needs to be democratized is where we should be.... It's not the boxes that deliver [media content] that [are] important [for our mission], but the idea of community media. [14]

In his identification of the group's priorities, he indicated that the particular technologies favored by the group were subject to negotiation. As one activist stated, "I love radio, but it's not going to be the same in ten years."[15] The activists' concerns were navigated in the midst of a shifting technopolitical media landscape; his affection for radio itself was not necessarily sufficient to support organizing around it over the long term.

Interpreting the Internet: Practices and Policies

The Telecommunications Act of 1996 ushered in a massive wave of consolidation amongst media companies. A primary rationale for consolidation had to do with the supposed availability of new media, mainly the Internet. Some regulators and broadcasters believed that traditional media would be subject to greater economic threats by new media, which necessitated the merging of the old guards. According to media historian and Free Press founder Robert McChesney, new media were seen as heralds of a more democratic media landscape. Many claimed that "the Internet ends the problem of broadcast scarcity (that is, more people want to broadcast than there is space on the airwaves) and means that everyone communicates on a relatively equal playing field."[16] With new media such as the Internet held up as harbingers of democracy, regulations to protect the public interest with regard to traditional media were valued less.

Due to the 1996 act's chilling effect on radio ownership in particular (see chapter 1), many people concerned about consolidation stepped up advocacy for legal access to LPFM radio stations during this period. At the same time, a multitude of unlicensed micro-broadcasters defied the FCC. As many as one thousand of these unlicensed broadcasters were on the air in 1997.[17] During this period, microradio advocates considered the possibility of using the Internet for "webcasting," but many did not view it as an equivalent alternative to FM. Jasper routinely argued, "If Clear Channel [a large corporate owner of radio stations] wanted to trade me my website for their 1,200 radio stations, I'd do it tomorrow.... A lot of people think we're crazy for focusing on this dinosaur technology, that some new pie-in-the-sky technology will come along and eclipse everything ... but people didn't expect radio to last after 1950."[18]

Activists had numerous objections to webcasting. Webcasting was an undesirable alternative to FM because it was less accessible. Internet connectivity and computer literacy were required to produce or receive webcasts. Speaking in 2003, one Philadelphia activist compared some available media choices, including low- and full-power FM, television, and webcasting:

> Radio is the most universally accessible ... you can reach more people. [Radio isn't expensive], you set it up, you talk into it, people tune in, on their car radios or whatever. The cost of entry of tv is just too high....
>
> We're looking at broadcasting to people who are within a mile or two of our studio; it is mostly our friends and neighbors. And that's probably one difference between a small community radio station and a big commercial FM station that covers multiple cities or at least a whole city.... So we're trying to come up with something that will be meaningful for that [nearby] group of people. With webcasting, you're trying to come up with something that is somehow unique or different, so you can somehow differentiate yourself from the other thousands of websites that are out there, something that will be meaningful to a group of people that are geographically distributed all over the world.[19]

Although the activists more routinely compared radio to Internet-based technologies, it is worth noting that this person also considered (and rejected) television. Activists also held that using computers to transmit or receive "broadcasts" was far more expensive than FM. And webcasting lacked the crucial element of "localism" that advocates saw as a main advantage of FM broadcasting. Another activist said, "Everybody has a radio, not everybody has a computer. You don't need any skill at all to be able to turn on the radio. Webcasting, anybody can get it, you can be in Oslo and listen to West Philadelphia radio. In a way, that's cool, if I'm a West Philadelphian in Oslo, that's great, [but] it makes it less somehow cohesive for the community.... If you webcast, it doesn't seem like a community resource anymore."[20] For her, accessibility and community orientation made FM radio more desirable than webcasting. Another radio activist said that FM possessed immediacy and even serendipity that webcasting lacked. "There's a sense of urgency about every minute. You're on-air, you know, it's happening ... [With the web], it's not broadcasting. There is some sense about [radio] being broadcast, you can turn on anything that's there, [in a] specific location, you can run into it by accident, but to go to a website, you have to know where to go."[21] Activists consistently raised concerns regarding "localism" or "community," accessibility, and use patterns that differed between FM and webcasting. They clearly favored FM.

However, this did not mean that people who preferred FM radio ignored the Internet entirely. Prometheus organizer Brian said in 2006, "We use 'radio' narrowly to mean FM broadcast radio, but it's more. You can have the Internet [connection] and a mesh network and tie them into the radio station.... [Y]ou could utilize the mesh network for production and uploading ... you could produce PSAs [public service announcements] at home and never have to go into the studio."[22] He also described how a wi-fi link could be used in the setup of an FM radio station to establish the studio-transmitter link. In this configuration, a wireless Internet connection would be used to send audio from the production site to the transmitter and antenna (an arrangement that Prometheus used in some LPFM stations) (see figure 7.1).

Using the Internet to share audio content for broadcasting was also a common practice within radical media. Jesse Walker wrote in 2001, "The A-Infos Project, a collective of online anarchists, has set up a website through which [unlicensed] micro stations, legal community radio stations, and independent producers can upload and download news reports, full-length documentaries, and other shows in MP3 form [digital audio format]."[23] There were other similar examples. Prometheus actually recommended that LPFM stations web-stream their FM broadcasts if they had the technical and economic[24] capacity to do so. But one activist said in 2005 that "not as many LPFMs are webcasting as you might think."[25] The radio activists did not entirely reject the use of the Internet for community media. Rather, they were critical of the suggestion that webcasting might be an analogous or comparable substitution for FM broadcasting. They were open to using the Internet to extend the practices of producing "traditional" FM radio.

Wi-fi networks and software-defined radio (or "smart radio") also present opportunities to examine the radio activists' attitudes toward the Internet's

Figure 7.1
Volunteers mount a wi-fi dish for a studio-transmitter link atop a church, Philadelphia (2008). Volunteer photo.

role in broadcasting and community media. Short-range transmission between wireless devices in portions of spectrum designated for unlicensed use is permitted by the FCC. These devices are commonly referred to as "part 15 devices," after the section of FCC statute governing them.[26] Part 15 devices include everything from extremely low-power FM radio transmitters (250 μV/m at a distance of three meters from the antenna, often understood as using a transmitter powered by around 1/25 watt) to wi-fi cards and baby monitors.[27] Wi-fi (or wireless broadband Internet connectivity[28]) is a later technical innovation that, similar to garage-door openers, cordless phones, and baby monitors, uses RF to allow devices to communicate across short distances.[29] Smart radio refers to the use of this spectrum by devices configured with software to change between dynamic frequencies to transmit and receive. This capacity to change between available frequencies has the potential to enable many more channels of communication in the same amount of spectrum. Devices only need to know with which other device they are communicating and select together the frequency to use. This potentially obviates the need for clear channels. One way to explain this is the metaphor of many people communicating in a crowded room; if two people sitting next to each other can agree to pay attention to one another, they can hear each other. Multiple whispered conversations can simultaneously occur in the same room. By contrast, a clear channel model is likened to one person speaking loudly in a room to an audience, permitting only one "conversation" at a time. Smart radio technology can be used to run wi-fi networks that dynamically change frequencies to rout around obstacles and communicate bi-directionally (transmit and receive) in order to network between computer users and share Internet service. This is the application that most excited the activists.

Wi-fi networks may be configured in a number of different ways. They range from open, nonproprietary, dynamic meshed networks to static, closed, proprietary hub-and-spoke networks (with other models in between). They are not inherently open to other users or devices not specified by the network.[30] The term *community wireless network* indicates "open, freely accessible, nonproprietary systems ... built using the buying power and economies of scale within neighborhoods, towns, and cities."[31] A "municipal" wireless network has a slightly different valence than a "community" wireless network. It indicates that the service is being provided or hosted by a municipality, as opposed to a corporate provider. But it does not necessarily connote the openness of a community network, and it may not embrace the use of the network as a platform for community media. Instead, these networks often provide users with connectivity that is largely similar to connectivity that would be provided by a corporation (these distinctions are discussed more fully below). It is also worth noting that not all community or municipal broadband efforts are also wireless—some are cable-based, though increasingly the model is to incorporate wi-fi.

During the period from 2000–2011, LPFMs were virtually impossible to license in cities (due to the requirements for spacing between stations that Congress placed on LPFMs in 2000). During LPFM's first decade, such stations were almost completely out of reach in urban areas. Activists also acknowledged that even if the FCC and Congress were to reauthorize the FCC's initial recommendations for LPFM (as finally occurred in 2011), LPFM licenses would still remain elusive for many in urban areas because of spectrum crowding. Partly due to the unavailability of LPFM as an option, the Prometheus activists considered the expedidency of municipal and community wi-fi networks in cities. Their interest was partly symbolic. Wi-fi could allow Prometheus to stake a claim in cities that would complement their work building radio stations in rural areas. One activist stated, "We care about radio, but we believe in appropriate technology.[32] If wireless is the best way to support community and social justice needs, we need to get them that. If people can't get radio, they need this now instead."[33]

The idea that the airwaves belong to the people was first articulated about radio broadcasting. But it gradually enabled the radio activists to claim community wi-fi as relevant as well. They were aware that the technical and political terrain on which their LPFM work occurred was constantly shifting, and they assumed that their focus might not be solely on LPFM as time went on. Their understandings of technology, political organizing, and policy work led them to actively imagine other ways in which their mission to promote a democratic media environment could be expanded from their more narrow focus on radio. One Prometheus volunteer looked back on the struggle for LPFM and said, "I was so stupid [then]. I thought that when [then-FCC chairman] Kennard said he was going to start giving out licenses, I thought this meant we had won."[34] She described researching telecommunications technology and policy issues in order to more fully understand the wider realm in which the group's work was situated. (As a single mother who had dropped out of college, she vividly recalled "reading Harvard economists' reports" while her toddler "peed in her lap.") This led her to the issues of media ownership and spectrum management, including open spectrum and wi-fi. She began to formulate ideas about how to expand Prometheus's scope, which included the idea of what the "spectrum" is in the first place. She said that for her, "Spectrum isn't a thing—the first step is you need to make it a thing."[35] Here she underscored that the notion that private property (as opposed to its technical aspects) dictated how spectrum is treated in the realm of policy: "Policy, more than science, informs our understanding."[36] She reiterated this in a presentation about spectrum management: "Building our own communications infrastructure today is technically possibly and economically feasible. It's not a pie-in-the-sky situation, the only obstacles are political.... Neighbors helping neighbors works better than the big companies."[37] The activists thus asserted that it was possible and desirable for communities to build wireless networks, expanding a position they had long maintained about radio to include other technologies.[38]

In 2004, Prometheus worked with other advocacy groups on a state legislative campaign in Pennsylvania. They mobilized to oppose a bill that would give Verizon, a massive telecommunications corporation, the right of first refusal before municipalities could set up their own broadband networks. The bill ultimately passed in late November 2004. It contained a clause that denied Verizon's right to challenge the city of Philadelphia's plan to build a wi-fi network, which the city government and other groups, including media activists, were already planning. Philadelphia's planned network was grandfathered in, but the ban was put in place for the rest of Pennsylvania.

By mid-2007, fifteen states had passed similar legislation banning municipal broadband initiatives unless local telecommunications corporations approved them.[39] Controversy over municipal wi-fi resulted in large corporate entities opposing each other. Providers of broadband service and content such as Comcast and Verizon opposed municipal broadband efforts. But makers of computer and networking hardware generally favored municipal broadband under the assumption that there would be greater demand for their products no matter who provided connectivity. A pilot wi-fi program in Philadelphia relied on donations from Cisco. Another program in Chicago received funding and hardware from Dell.[40] Prometheus was interested in municipal wireless for a number of reasons, including the strategic importance of organizing in solidarity with other media advocates and consumer protection groups.

Prometheus members had also developed an interest in wi-fi as it related to their own mission. The activists emphasized material links between wi-fi and FM radio. The notion of the spectrum was crucial—symbolically, politically, and materially. Similar to FM radio, wi-fi also uses RF as the technical means by which data signals are transmitted. This material continuity between wi-fi and radio was often used to explain what wi-fi is and how it works. Representations of wi-fi often drew on radio as a familiar technology to indicate how the newer technology should be understood. Many illustrations of wi-fi "hotspots" showed RF radiating from what looked like radio towers.

One of the activists' intents in challenging dominant media institutions was to provide a platform for the creation of alternative discourses. They favored media "content" produced by ordinary citizens that could stand in contrast to the content provided by commercial media outlets. Esmé reflected in an interview:

> All kinds of technology, and I think especially communications technology, creates leverage and power. Often, the thing [our society] understands best to do with [technology] is to oppress people.... People that need communication the most have the least access to it. You can almost define oppression by lack of ability to communicate, to express yourself, to be heard, to be able to bear witness to your life, to be able to network with other people, to be able to create agency [for] change, to be able to celebrate your culture.[41]

Critically, activists held that media technology could empower people along two interrelated lines. It could enable them to "tell their own stories" and express themselves. It could also give them the means to challenge elitist and technocratic decision making.

A main strategy that Prometheus employed was to teach people to build and use technical artifacts. As explored in previous chapters, activists hoped to teach people to be unafraid of technology and to challenge expertise. They felt that if people attained technical competence, they would extend this sense of agency to nontechnical matters and become critical of expert-based, technocratic decision making. The propagation of technologies was an important part of Prometheus's vision. Ellen commented that she found the material and symbolic value of wi-fi to be complementary to her radio activism agenda. She discussed the organizing strategy of holding workshops in which coffee cans were used to build directional antennas for use in wi-fi networks ("cantennas"):

> The cantennas [are] an organizing tactic. It's an easy piece of technology to build. It's a useful piece of technology. In the ten or twenty minutes it takes someone to learn to use a cantenna, you learn RF, you learn DIY sharing of a public resource, like public airwaves stuff, you handle a drill, you handle a soldering iron, you have them handle a component, you learn about cabling, it's a fucking barnraising in a ten-minute package; it's the best tool for that.[42]

She made reference to the material linkage to radio when she said that a cantenna workshop teaches people "about RF." The connection to radio is also apparent in teaching people to use soldering irons and become familiar with cabling while building cantennas. But the main significance of the cantenna workshop for her was that it was "like a barnraising" in that it combined Prometheus's technical and political missions, raising awareness about citizen use and ownership of the spectrum. The activist also hailed the cantenna workshop as an effective organizing tool because it was more portable and less involved than a barnraising. As a result, it could be shared with more people (see figure 7.2).

Incorporating wi-fi into their organizing mission proved troublesome for Prometheus. The activists' understanding of radio as ideally suited to demystification of technical expertise did not necessarily translate to other artifacts. One Prometheus organizer stated, "With radio, it's easier to have a real 'Eureka!' moment, like when you realizing you're broadcasting from a [radio transmitter mounted inside a lunchbox]. This is harder with computers."[43] A Prometheus intern echoed this: "The barrier [of] access to radio is so much lower. You have to know relatively little [technically] to produce or use radio creatively, but with computers it's much higher."[44] In previous chapters, I described how soldering a transmitter board was a

Figure 7.2
A finished cantenna to be used in a studio-transmitter link or on its own in a wi-fi network. Volunteer photo.

good opportunity for novice participation. It is relatively simple to solder together the various components, as long as the instructions and schematic are closely followed. It takes several hours of work and is a social activity. Each small board can accommodate a couple of people soldering and at least a few more observing or guiding. When the board is complete, it is easy to hook it up and demonstrate its use to broadcast an audio signal. By contrast, a cantenna has a less obvious function when it is complete. It is an artifact that can be integrated into a network of other technologies in order to produce a wi-fi signal and link computers. But the computers themselves are still complex and essentially black-boxed. The cantenna is arguably a more abstruse end product than a voice or music sample being heard over speakers.[45] This is not because of any inherent properties of these artifacts. It is because the stabilization of the use and meaning of radio as an artifact enables a transmitter's function to be readily grasped by novices.

Radio's common understanding as a medium of sound transmission may also make it more easily understood as having democratizing implications. With radio, the idea of a "voice" is salient. It is not only an aural phenomenon. It also resonates with the ideas of "having a voice" or "being heard" within discourse surrounding democratic participation.[46] To illustrate this point, one activist stated that barnraisings were part of an "international movement for people to own their own voices";[47] *voice* is about power.

Activists were stymied in their efforts to make wi-fi seem as transparent, utilitarian, and democratic as radio. Although they largely understood it in these terms themselves, they had difficulty convincing the public that their robust vision for community wi-fi was about more than Internet connectivity. Activists wished to promote wi-fi for the circulation of locally produced media, but other uses of the Internet had already stabilized. The dominant understandings of connectivity inhibited the activists' efforts to symbolically forge a link between wirelessly networked computers and the ideas that had crystallized around radio. Radio is not inherently a more democratic technology than wi-fi. But the radio activists were swimming upstream in their attempts to promote an interpretation of community wi-fi that diverged from how Internet connectivity was commonly understood. This illustrates the wider point that technological meaning has to be constructed. The political valence of a given artifact is a result of interpretive work.

Technological Mediation and Its Discontents

In promoting their preferred meaning for wi-fi, the radio activists faced dilemmas in multiple directions. Would-be users did not always value wi-fi in the terms they did. Nor did policy makers or other geeks. The radio activists also struggled with issues of position. When grassroots demand for technology in underprivileged communities was at odds with their own understanding of technology, they exhibited discomfort about paternalism. These dynamics can be observed by attending closely to the use and meaning of wi-fi networks as promoted by different groups.

Municipalities and nonprofit groups often cited reasons for building wi-fi networks that did not identically match the activists' interest in wi-fi. In early 2005, a Prometheus organizer met with a representative from a nonprofit group in Philadelphia called HousingSpace (a pseudonym). The organization was a former homeless shelter that provided other services such as computer access and job training. It had built a wi-fi network in the neighborhood in which it was located. In the meeting, the HousingSpace staff member stressed the use of the wi-fi network for services such as downloading forms from city social services agencies. In many areas, including the one in which HousingSpace was located, new wi-fi networks also required the provision of personal computers and training to use them. Many families and individuals receiving wi-fi access had never before owned computers. So-called digital inclusion was a complex process, confronting historical exclusion that exceeded "the digital."[48]

This emphasis on basic computer literacy and the use of connectivity to primarily *download* material (or to eventually use connectivity for purposes such as commerce or running one's own business website) represented a paucity of vision as far as the activists were concerned. It was also paternalistic: lawmakers and some nonprofit organizations tended to represent the "users" as wards of the state. During a 2006 city council session, one Philadelphia City Council member said that the benefits of a municipal wireless network would be to provide "high speed Internet to all citizens and businesses, to take advantage of the new digital society. [We can] bridge the digital divide in 12-18 months, provide access and opportunity for all, prepare children for the future, empower low-income families by providing access to information and social services at home, [and] level the playing field for small businesses."[49] By constrast, Prometheus activists and others who favored community wi-fi saw the potential use of these networks as extending beyond the provision of Internet service. Instead; their interest flowed from their vision of wi-fi networks as *platforms for community media.* Significantly, they emphasized *uploading* content and multidirectional transmission as opposed to downloading news, entertainment, or forms related to services. One document distributed by Prometheus stated,

> People just like you have been using inexpensive wireless transmitters to shoot high-speed internet from home to home and neighborhood to neighborhood.... They've expanded wireless networking from a way to get the tangles of cables out of your home office to a way for communities to get the connectivity they need for cheap or free. In some cases—like right here in [this town]—they are *redefining the internet altogether!*[50]

To the radio activists, community wi-fi was appealing due to its potential for unrestricted and multidirectional transmission of citizen-created content. One e-mail sent by a Prometheus organizer expressed her concerns about the city of Philadelphia's plan, which had not yet been fully outlined: "will the important community content—like the videos produced at [a community] video center, the content hosted at the IMC [Independent Media Center], and the community newspapers and websites scattered across the city—be marginalized or promoted to users of the network[?]"[51] Advocates touted community wi-fi as "cheaper, more reliable and flexible, and offer[ing] end users access to more bandwidth, services, and applications" than profit-driven corporate models.[52] They added that participants in a community wireless network could decide to create such resources as streaming media servers. This possibility represented a major difference from a model in which users of broadband were assumed to be primarily or exclusively "consumers." A document prepared by a Chicago nonprofit with whom Prometheus consulted echoed this interpretation of wi-fi: "It's important to understand that a connection to the Internet is just one of the many services a [wireless community network (WCN)] provides. Because a WCN creates a very high-speed network local to your neighborhood, you'll be able to receive interesting content that your community produces while sharing content that you produce ... The WCN achieves speeds higher or comparable to DSL or cable modem. Additionally, a WCN is community-based and delivers content and applications that are community-created and community-specific."[53] In a document prepared by Philadelphia activists with whom Prometheus often collaborated, this sentiment was expressed even more strongly: "Communities across Philadelphia are *fighting to tell their own stories.* The city's wireless plan could give thousands of us a new way to do just that, but we need to let the city know that, when it comes to technology, the public interest is the criteria [*sic*] for success."[54] A Prometheus organizer stated, "It's time to take back unlicensed airwaves—wireless community networks are not to just receive content but to create and transmit it."[55] The act of *transmission* was crucial.

Due to their interest in community wi-fi and their belief that it was an "appropriate" option in cities, Prometheus consulted on a project with a Chicago-based nonprofit organization, Neighbors for Access to Technology (NAT) (a pseudonym). NAT had built a small wireless network consisting of a few nodes, which were used by a neighborhood community center and a few homes. NAT was planning to expand this into a larger community wireless network in the economically disadvantaged, largely African American Chicago neighborhood of Larch Park. Their staff were interested in collaborating with Prometheus on this project, especially hoping to draw on Prometheus's expertise in leading hands-on workshops in the mode of barnraisings. But after arriving in Chicago for meetings, the Prometheus activists began to feel uneasy. Prometheus tried to get a feel for the reaction of Larch Park residents to the proposed wi-fi network. Some were enthusiastic. A person who worked closely with the community members said that "this is a chance [for the residents of Larch Park] to not just keep up with society, but to advance beyond it; people want to use this to start businesses and for education."[56] In this comment, he alluded to the historic exclusion of African Americans from technological decision making and "progress."[57]

However, the Prometheus organizers were concerned that this goal on the part of the residents stopped short of their own agenda, which included social change and local citizen–created media, not only connectivity.[58] They also had reservations about collaborating with NAT, an organization they felt differed from Prometheus in significant ways. One Prometheus activist later characterized another organizer's reaction, saying, "[H]e didn't like the [NAT] people ... he didn't trust them, he didn't like the money that was there."[59] A Prometheus intern privately referred to the situation with NAT's attempt to bring Prometheus onto the community wi-fi project as "a liberal clusterfuck." He felt that NAT had good intentions, but the project was not well managed. He worried that NAT exhibited an attitude that could be construed as heavy-handed or patronizing toward Larch Park's residents—an attitude from which Prometheus members wished to distance themselves.[60]

This project involved money and technology coming in from outside the community. The neighborhood residents had a (legitimate) concern that this sort of investment in improving the neighborhood might not solely be in the interest of the current residents. A historically poor neighborhood in a desirable location with well-developed communications infrastructure might be a target for gentrification, for example. There are historical reasons why African Americans in some cases may have an adversarial relationship to technology, particularly that introduced by whites.[61] Some residents raised concern over attempts by a group of (largely white) people from elsewhere (NAT was not based Larch Park, and Prometheus was not even from Chicago) to "improve" the neighborhood. A NAT staff member said, "There will never be a time when it will be okay for hundreds of people who don't live in [Larch Park] to come volunteer there."[62] This made Larch Park a troublesome site for a wireless barnraising. Organizers concluded that Larch Park was not an appropriate site to promote Prometheus's wi-fi and general open-spectrum interests. Ultimately, Prometheus participated in a smaller project to build network nodes in the neighborhood. In order to support their community media agenda, they worked with a community member to get a grant for equipment to start a community Internet radio station for the neighborhood (over their wi-fi network as opposed to FM). They did not conduct a wi-fi barnraising in Larch Park, nor did they conduct a stand-alone wireless barnraising elsewhere.

Prometheus organizers were concerned about Larch Park because they did not want to project an image as (white) paternalists. For the activists, providing nonwealthy citizens and community groups with radio stations was another means of leveling power and promoting egalitarianism and pluralism. The activists' vision promoted the inclusion of as many groups as possible, including (or especially) those lacking some forms of social or economic capital. Nonetheless, this was a difficult area for the Prometheus organizers. Though they were critical of white privilege and paternalism, their organization and volunteers were made up of an educated, white[63] segment of the population. The activists were therefore extremely careful to frame their activities as self-consciously antiracist, promoting cultural exchange. They did not want to be seen as a group of (mostly) white activists providing a commodity or service to a less-privileged "other." In an interview, one Prometheus organizer commented about their international workbuilding radio stations in Nepal, Tanzania, Kenya, and Guatemala:

> Historically the US has been seen as a patronizing force, NGOs come in.... But we're learning from [community groups] in other countries—we're resource-rich materially, but how are they organizing? We want to find winning strategies and learn from them and use that knowledge here [in the United States or Global North;] they are doing community organizing under [circumstances that are unbelievable]. We're not "giving" to them, because in the US, the organizing strategies and ability is in its infancy.[64]

He also stated that "privilege allows us to not realize that [media] is a life and death issue for other people. As a white[65] activist group, we're in solidarity—they can use the radio station to do it themselves."[66] This is not to suggest that the activists succeeded in evading charges of paternalism in their organizing activities. It is only to highlight that they expended significant effort reflecting on these issues.

Another organizer commented on the potential tensions for Prometheus in choosing groups to hold barnraisings with: "It's easy for us to work in rural communities where we're a big deal when we come in, where they want us to be there. To be honest, it's easy to work with other nonprofits. It's easy for us to work with other white groups.... We can work well with hippies. We've done a great job of it in the past."[67] In noting that that Prometheus worked well with "hippies" and nonprofits, this activist was remarkably candid about the ease in collaborating with groups positioned similarly to themselves in terms of race and class. Prometheus's identification with hippies (and vice versa) was a legacy of their countercultural heritage and communalist ideals, as discussed in previous chapters.

She added, "I'm impressed that we've been able to work with farmworker groups. I think that that shows a lot of growth...,"[68] indicating that the group strove to break out of their comfort zone when selecting collaborators (see figure 7.3). The activists' attention to "difference" and its potential to stir up tensions occurred in their work with LPFM and in their work with wi-fi in Larch Park. Another activist said that privately, the group struggled with how much to foreground issues of race and class in their work. He said that occasional experiences forced the group to reflect on these topics and spurred internal dialogue within the organization. He felt this was positive, even when the conversations were hard or forced them to confront ways in which the organization needed to improve.[69]

The Prometheus organizers' experience with the Larch Park case was complicated. But critically, their status as a white group providing services to a group with whom there was a perception of social difference was not the only concern. Although this was a potentially vexing issue on its own, it was one with which Prometheus routinely grappled in their work building radio stations domestically and internationally. (I have outlined above the resources on which Prometheus members drew to legitimate and explain their encounters with groups who were "different" from them.) In Larch Park, race and class differences were significantly compounded by the fact that the grassroots demand for community wi-fi *as it was understood and*

Figure 7.3
A barnraising workspace inside a farmworkers' union, Oregon (2006). The mural reads: "*¡Respecto y sueldo justo para los campesinos!* [Respect and fair wages for farmworkers!]." Author photo.

promoted by Prometheus was absent. Larch Park residents desired *connectivity*, but did not flesh out uses for a wi-fi network beyond "education, business, and keeping up with society."

Prometheus members believed themselves to have a strong grassroots mandate. They constructed an organizational identity around this belief. The discomfort Prometheans registered about NAT could be read as attempts to embody a "different" kind of nonprofit organization (as explored in chapter 5). In their board meeting, the radio activists reflected on these issues. Even though they felt that they had compelling reasons to enfold a spectrum management agenda into their mission (based on their own understanding of the technical and political issues at stake), they did not see how they could do so without perceiving a grassroots mandate to support this work. They felt obliged to listen to "the voices of the people." They did not want to "carpetbag organize." (Again, I am not suggesting we take these claims at face value, only that they were important in the activists' ideations about their positions and their organization.) One of the activists captured their dilemma, saying, "No one goes around with signs that say 'Free the Spectrum!'"—and another chimed in, "Except us!"[70]

Activists worried that wi-fi was a bloodless issue: no one was ready to "fall on a bayonet" or go to jail for wireless. Prometheus held passionate opinions about wi-fi and spectrum management, but activists could not assume that members of the public shared their sentiments. Without outreach and education efforts, the grassroots demand for community wi-fi was less easily identifiable than the appreciable demand for radio stations. Prometheus hoped to differentiate itself from "wonk," "Beltway," or paternalistic organizations, so the lack of grassroots demand was troubling. Even when grassroots demand for wi-fi was apparent, it could not be assumed to be for wi-fi as the activists understood it (as opposed to a general demand for Internet connectivity). As one activist said, "we can't push [our agenda] on anyone. We're patient because we have to be; we can't do things before groups are ready."[71]

The "wonkiness" of wi-fi troubled the activists. They were often critical of other groups whose purposive engagement with technology did not, in their opinion, lead to a more egalitarian distribution of expertise. The radio activists were critical of computer hackers and free and open source software (FOSS) developers. They felt that these people tended to not be politicized enough. Their knowledge was too elite (as described in chapter 2). When advising an LPFM station about software, a Prometheus activist recommended they avoid an open source platform unless many people at the station were already familiar with it. He said, "The problem with using open source is that it puts [the radio station's operating system] more into the hands of the cadre of nerds—it's not a platform that as many people know, which, practically, is a problem, even though politically it's great."[72] Another activist who was involved in radio activism and open source reflected about the radio activists' engagement with, and sometimes opposition to, hackers and FOSS developers:[73]

> The open source community has potential to be activated in a political way [but it isn't always]. Open source software is free in a number of different senses, it's free in that anyone can use it freely, it's open, it's free for anyone to go through and change, it's free in the freedom sense, in that some of it, you can use it and change it, but whatever you put into it has to remain open and free. So it contributes to a general culture of people contributing to a general pot of common tools and common infrastructure, and they do it for free....
>
> [There's] a general sense that ... these technological tools are powerful and should be used for a common good and not be developed for profit when profit is at the cost of human need, [which] is exciting and has a good intersection with a lot of the values of progressive and social justice movements....
>
> So open source projects that also have a political focus are really exciting.... [F]ocusing that energy on more socially useful open source software, I think would be great.[74]

This radio activist had thoughtfully considered the ways in which the FOSS community could be brought into line with more overtly political goals, building technological tools for an explicit social change agenda. But he drew a distinction between FOSS as a general project and FOSS projects with a deliberate political focus, suggesting that FOSS participation in general was not consonant with the radio activists' rather more robust politics. He believed hackers were related to the radio activist mission but distinct from it. He imagined that things would remain that way until or unless hackers took a more sensitive political stance regarding their technical engagement:

> The other one that's kind of interesting is the hacker community, which is kind of related to microradio stuff.... [W]e kind of want to court hackers.... People get involved in [technical projects] for a lot of reasons, [such as] they're bored, they're disillusioned, they have technical interests and they're isolated and there's something about their interest that is not being satisfied ... [but] there's often a lack of a goal....
>
> We try to maybe draw them out in such a way that they have to figure out what their values and goals are, and then maybe involve them in more political work, on something that's a little more focused on the community beyond the technical community. That seems pretty exciting, and pretty possible.[75]

The radio activists were keen to reflect on ways in which they felt that their work and goals resembled (as well as differed from) the work and goals of members of related technical communities such as hackers and FOSS developers. They tended to view their own work as more well-developed politically because it included an overt challenge to elitism in technical practice. They wished to further politicize members of hacker and FOSS communities and inspire them to work toward more fully articulated activist goals. Radio activists viewed the FOSS community as overly concerned with building technical artifacts at the expense of awareness of the potential problems created by elite technical knowledge. This was especially a concern with issues related to the inclusion or exclusion of novice participants. (Chapter 4 addresses the potential gap between "participatory culture" and the radio activists' desire to promote egalitarian technical participation.) (These critiques did not preclude the radio activists seeking FOSS developers' or hackers' help when they needed assistance in projects requiring software development and other technical matters.)

The FOSS community was also mediating wi-fi technology, developing material artifacts and resources for interpreting these artifacts. Users of open platforms such as Linux have been involved in "user-driven innovation" in the development of wi-fi networks,[76] not unlike the creators of the ARPANET (the progenitor of the Internet).[77] A document distributed by Prometheus activists stated, "Wireless networking by the community of geeks and experimenters who have been innovating networks on the cheap for years, [this website is] much more focused on cheap, free software-based and open-source solutions for connecting homes to each other wirelessly."[78] By 2004, all major wi-fi card manufacturers released had drivers for Linux, which illustrates the relevance of FOSS communities in developing wi-fi (and especially the card manufacturers' interest in courting this market).[79]

Material and social links between the FOSS community and those building wi-fi networks were strong. The radio activists' deepening interest in wi-fi brought them (by necessity) into even closer contact with people involved in FOSS projects. Radio activists' interest in wi-fi was generated in part by the involvement of people they knew. Indeed, it is inaccurate to treat these as wholly discrete social groups. Some members of each group were involved in telecommunications policy, programming, and technical work. Members of both groups attended such events as HOPE (Hackers on Planet Earth) conferences and barnraisings. Radio activists possessed some degree of overlap with FOSS developers in terms of interest and skill in FOSS, programming, or computer hardware; they also expressed some major points of contention, as previously noted. The FOSS community's approach to building and promoting wi-fi gave the radio activists resources for understanding wi-fi. It also created difficulty when the radio activists' notions about how to build and promote wi-fi clashed with those of FOSS developers. Prometheus continually cited difficulty with wi-fi promoters, claiming that they were "too wonky." They also claimed they were inattentive to the grassroots and to wider issues of social justice.[80] Collaboration was further hampered by the fact that other geeks did not share the radio activists' vision of radically egalitarian technical participation.[81]

Pondering Prometheus's foray into wi-fi, one organizer stated that "the radio stuff is very tangible. People learn and then they are passionate. Our organization is the whole package for radio, but for wireless [we have had less success]."[82] This comment reflects some of the difficulties the group had in finding the right way to promote wireless as an accessible and desirable technology. It also underscores the observation that working with other groups on media democracy and spectrum management issues was a challenge. One regular feature of their organizing work was to consider groups with whom they could ally. They often exchanged support on wireless issues as an in-kind trade with groups who could make an impact on LPFM-related campaigns. In this respect, wi-fi was also a means to an end in terms of promoting LPFM. The radio activists felt that relationships formed with other advocacy groups or legislators could be leveraged in a variety of ways.[83] But they were sometimes left with the lingering feeling that other organizations would potentially be better equipped to combine the technical work and the "message" needed for a successful campaign around wi-fi. Prometheus's formula for success with FM was not easily ported to community wireless. Despite this, groups who valued the work Prometheus did with FM were interested in tapping into Prometheus's ease with hand-on demonstrations concerning the material and political aspects of communications technology into wireless. This was why they had been asked to join the effort in Larch Park in the first place.

Conclusion: The Interplay of Old and Emerging Technologies

In distancing themselves from groups that more unabashedly embraced digital technologies and the Internet, the radio activists provide a unique site for analyzing new media adoption and resistance. They are not dismissible as mere Luddites or nostalgic radio hobbyists. In fact, their high profile in the media democracy movement indicates that they were taken seriously by advocacy peers. The radio activists' attitudes toward webstreaming and community wi-fi demonstrate that the negotiation of new technology can have subtle contours; adoption and resistance of technology occur along a continuum. Over time Prometheus cautiously expanded its definition of "appropriate technology" in order to include community wi-fi networks. But what radio activists valued in the newer technology was heavily informed by what they elevated in LPFM—in particular, the ability to locally "broadcast" citizen-created content. In this, they preserved the notion of user agency they imported directly from their understanding of FM radio.

The truism that new technologies necessarily "disrupt" and overthrow past practices is a distortion. It is not an accurate depiction of the more gradual and evolutionary events that occur on the ground with the introduction of new technologies. New technologies' salient and "unique" qualities usually begin with borrowing from social practices that surround existing technologies.[84] In this case, the activists did not advocate the acceptance of new technologies until they could locate and articulate continuities between radio and community wireless networks. By pointing this out, it is not my intent to characterize the radio activists as especially savvy or prescient. But these dynamics do tell us something about the trajectory of new technologies. They also accentuate the importance of leaving "old" technologies in the mix as we assess new ones. We do not need to know how these artifacts or wider issues about community media will "settle" in order to derive meaningful insights from this case. Listening to the uncertain early stages of these negotiations can help us understand the trajectory of technological change without succumbing to the hype of sudden, marked, revolutionary change (which is usually technologically deterministic).

In their capacity as mediators, radio activists had the potential to shape how users understood or interacted with certain technologies. Through advocacy work, they also had the ability to influence whether citizens would have legal access to certain technologies, including FM radio and community-municipal wireless.[85] This mediating role was at times uncomfortable, especially as they organized around community wi-fi. This was because of the perception that users and other social groups did not understand wi-fi in the same terms as the activists. Promoting community wi-fi networks as platforms for community media rather than Internet connectivity illustrated this dilemma. (By contrast, the grassroots demand for radio stations was largely consonant with the activists' vision for LPFM.) The notion of *broadcasting* was an enduring one, and this interpretation of community wi-fi provided a material and symbolic link between FM radio and community wi-fi. A document produced by Prometheus about wi-fi and smart radio (*not* FM) contained the statement, "Using a combination of the techniques outlined [here], it is possible to imagine a world in which anyone can be a broadcaster."[86]

At the same time, the radio activists were not restricted to advocacy. Their role as propagators meant that they labored to bring the same politics and values they had identified in radio to wi-fi through hands-on technical engagement. Computers are in some ways more opaque artifacts than radios. Their dominant meanings proved challenging for the radio activists to contest. Nonetheless, the activists tried hard to link their rhetorical interpretive work to technical practice. They promoted political and technical aspects together to users through cantenna workshops.

The radio activists' mediating role became additionally complex as it intersected with issues of race and paternalism. Ron Eglash et al. have explored the appropriation of technology by marginalized groups.[87] Here, the end users did not necessarily understand why they need a given technology in the same terms as the activists did. The activists had difficulty reconciling their belief that user groups "needed" technologies with their stated belief opposing (white) paternalism. Wi-fi proved vexing because even when groups presented a grassroots demand for it, this demand potentially "limited" to a desire for connectivity. It was problematic for the activists to creatively reclaim wi-fi on behalf of marginalized groups.

The radio activists' disconnect with users was mirrored by their disagreement with other mediators in certain ways. They were skeptical of municipal and NGO interventions that configured users as dependent. And they also disagreed with FOSS activists about how much of a premium to place on "the technical" versus "the political." One activist said, "There is a tendency among nerds to find technology to solve social problems. I would say it's just the opposite, you have to use the society to govern the technology."[88] Ironically, though, the radio activists did themselves elevate the technical when they promoted skill sharing in minibarnraisings and cantenna workshops. This may have unintentionally distanced them from would-be users, some of whom wanted "access" to communication technology without necessarily embracing the full suite of hands-on skill sharing the activists prized. Barnraising tactics could have hindered activists organizing with groups such as the Larch Park community group. The historical exclusion of African Americans from engineering culture and infrastructural development meant there were excellent reasons for neighborhood residents to be skeptical of white activists bearing technology coming from outside their community.[89]

Nonetheless, the activists found other ways to justify their interest in the Internet. They called wi-fi an "appropriate technology," in a rhetorical effort to bring it into alignment with other small- or community-scale technologies they favored. A board member voiced support for Prometheus's move into wireless, stating that "the Internet didn't drop down from the sky—it was created by the military and we need to take it back."[90] This interest in reclamation or seizure of technology indicates the influence of anarchist and Marxist traditions on the activists. On more than one occasion, the activists invoked the ideal of "seizing the means of production lest they be used against you."[91]

The radio activists sought to provide technical and symbolic links between their deep, loving, and playful engagement with radio technology and wi-fi technology. This aided them in their goal of teaching ordinary people relevant technical skills. Their hands-on work with technical artifacts like cantennas also provided them with an opportunity to create continuity between their identification with radio technology and their perhaps burgeoning identification with other technologies. But their rejection of digital utopianism made it difficult for them to develop the same unambiguous, affective relationships and strong identifications with computers that they had successfully cultivated with radio.[92] Although the activists could muster a strong enthusiasm for a portable transmitter screwed into a lunchbox, for a tool-belt, or for knitting, they would not extend this affection to a portable digital wireless communication device such as a smart phone. Of course, these meanings of technologies do not reside "inside" the technologies themselves—they require construction, maintenance, repair, and translation.

As media historian Carolyn Marvin writes, "New media, broadly understood to include the use of new communications technology for old or new purposes, new ways of using old technologies, and, in principle, all other possibilities of the exchange of social meaning, are always introduced into a pattern of tension created by the coexistence of old and new, which is far richer than any single medium that becomes a focus of interest because it is novel."[93] This episode in early-twenty-first-century radio activism exposes complex negotiations surrounding differing technological options (radio versus wi-fi). It also exposes negotiations to identify the best interpretations of a single technological option (wi-fi alone), some of which drew from interpretations of radio. This interplay demonstrates the continuing viability of an old communications technology (radio) and its centrality in understanding emerging technological options. Wi-fi's material connection to radio may have been partially responsible for the activists' interest in it. Most intriguingly, the radio activists only expanded their purview to include wi-fi when they could make an argument that it could be used in the same hands-on, empowering, and community-building ways that were possible with radio.

Conclusion

After a decade of advocacy, LPFM's proponents celebrated a victory. Congress at last passed the legislation they had pursued since 2000, and President Obama signed the Local Community Radio Act of 2010 into law. This law vastly expanded LPFM, allowing up to a thousand new small-scale, independent stations to be built nationwide. New stations will perhaps go on the air at about the same time this book is published, in 2014. Prometheus was a major force in shepherding the bill to its passage. The signing of this law marks the closure of the period discussed in this book. It is also a turning point for small-scale broadcasting. It would be tempting to speculate on the future of radio at this moment. Certainly, its demise is less foregone than detractors' death knells would indicate. FCC statistics released in early 2012 showed that the number of radio stations across most categories (AM and FM commercial stations and FM educational stations) had increased in recent months, and LPFM was of course poised to grow as well.[1]

It was never the intent of this book to prognosticate on the future of radio or even the future of LPFM. Nor was it my goal to offer policy prescriptions for community media or radio specifically. Although a number of media critics and scholars have persuasively argued that corporate consolidation of media harms political and cultural life through the narrowing of news reporting and centralization of media production, this book has not at any point engaged those claims directly.[2] It is plainly apparent that media outlets that are formulated as vehicles for accumulation of profit largely ignore other ideals of what a media system is meant to be. I have come to share the perspectives of media activists (and many scholars) on these issues; there is a pressing need for an independent, credible, and diverse media landscape (including media that are not profit driven). And yet, my aim is not to offer prescriptions for policy or prophesy about the fate of radio. I instead end this book with a discussion that is, in some ways, orthogonal to these issues. Still, my conclusions are vital to an understanding of the future of media technologies and institutions.

As cultural historian Thomas Streeter writes, "Sometimes exploring the complexity of what has actually happened offers more useful insight than the urgent gropings of prognostication."[3] What I hope I have provided in the preceding chapters is a nuanced ethnographic accounting of what actually happened: how actors vested an artifact with meaning, through a combination of technical practice, advocacy, and pedagogy. Although the radio activists' experience does not translate in "generalizable" ways, we can take away three concluding points from the dynamics explored in the book.

The radio activists bound together technology and politics in ways that make them distinct from activism that merely uses communication technology in the service of political organizing. The putative role of media and communication technologies in the organizing of political protest has been heavily scrutinized (most recently, in the context of the Arab Spring and Occupy Wall Street uprisings). But this book is concerned with a phenomenon in which media technology *itself* is the object of political activity. These activists not only diffused radio technology but also they actively built it. In the process, they sought to foster a democratic politics of expertise predicated on the sharing of technical know-how.

A main focus of this book is the activists' attempts to make technical practice and identity formation more inclusive. In this, they mounted a significant challenge to technical "participatory cultures" that take for granted democratic potential in self-organized projects. Many purportedly open technical cultures do not commit to the hard work often required to truly open up participation (either because their practitioner base is monolithically elite already or because commitment to nonhierarchy is more nominal). Yet regressive potentials can hide in emancipatory claims. In attempting to break the conventions of expertise as it is traditionally constructed in order to promote an egalitarian ideal, Prometheus ran afoul of real differences in knowledge and familiarity with electronics. In particular, the divide between novice participants and others deeply familiar with radio technology (including some with formal engineering training) was not easily overcome by a simple prescription to include novices or to disallow anyone from doing anything he or she already knew how to do. The skill and affective entrainment this proposition required could not be imparted over a weekend, as the transmitter workshop, Geek Group, and barnraisings showed.

Activism oriented around propagating technology highlights deep contradictions between egalitarian politics and technical cultures predicated on elite forms of practice. There are good reasons to question activists' romantic notions about the emancipatory potential of learning to tinker with electronics. It is also not clear whether their labor of love surrounding technology should be a universal goal.[4] Why should everyone want to have the skills to build a radio transmitter (or program a computer)? At the same time, this impulse to open up exclusionary technical cultures by hitching technical practice to radically inclusive politics offers a model of expertise that those prizing democratic participation may find compelling and useful.

The radio activists went further. Their claim was that "demystifying" technology—an especially abstruse and expert domain—was a socially empowering practice. It could politicize everyday people, leading them to question expertise in domains *exceeding* technology. (Though challenging expertise in technical domains is important in its own right, because technology is the seat of much cultural and social power.) This broad questioning of expert knowledge was at the core the radio activists' mission to "open up" technology. They conceived of technical skill and technical identity as universally attainable and appealing. It was not necessary to achieve expert status oneself. All that was required to develop this capacity to question social order was the ability to identify as a person with agency over technology. Technical identity was intended to function as a resource for the formation of broader political empowerment.

Yet there may be real risks in fetishizing technology as a platform for political action. This strategy may inadvertently reinscribe patterns of exclusion that have already formed around technical practice, limiting participation to those already inclined toward affective pleasure in technology. The radio activists' experience shows that these patterns of exclusion map onto gender, race, and class patterns stemming from the history of engineering and electronics tinkering. This is not surprising given that these practices have a long history of association with white middle-class masculinity. Exclusion was a profound and vexing issue for a form of activism predicated on technological affinity and egalitarian participation. The centrality of technical practice in this activist enterprise widened the gulf between the activists' stated ideals and what they were able to accomplish. Technical practice *itself* provided distinct challenges to those who sought to promote egalitarianism. This calls for a nuanced understanding of the symbolic and practical dimensions of placing technology at the core of an activist politics and praxis.

Although there are limitations to promoting a technical DIY ethos as a universalist project, we cannot dismiss technical practice as a platform for activist politics. The activists constructed technical expertise to mount a challenge to the notion that groups with less expertise should be excluded from technical decision making. However, equally distributed technical expertise may be the wrong focus for activism. The dictum that everyone should be an expert is arguably its own form of technocratic coercion (however well-intentioned). Why not reframe the conversation as one about social power instead? Elite social power and technical participation are imbricated to such an extent that they may at first glance seem interchangeable. But increasing participation in technology is no guarantee of movement into a more empowered social position.[5] Thinking about *multiple* axes along which people can be empowered is likely to yield better results than addressing narrow technical "divides" as if they were not part of much larger social and political configurations. (We might find that in a more equal society technical expertise is still not universally distributed. But this might not be such a problem if multiple avenues to opportunity and agency exist for all.)

A second concluding point of this book is to note that there is much more work to be done to understand the continuing development of old technologies. Activists' efforts to open up technical and political practice occurred in relation to radio, an old technology. The fact that the artifact around which activists oriented themselves was decades old makes this story interesting as a social study of technology in its own right. It is rare to have the opportunity to examine the reshaping of an established artifact so long after its introduction.

Communication historian Carolyn Marvin's *When Old Technologies Were New* might be followed by an inquiry into "when old technologies are old."[6] "Closure" for technical artifacts can be elusive, as acknowledged by scholars of science and technology studies. Often as a result of reinvention or user-driven innovation, technologies may not stay "settled."[7] Yet many studies of technology focus on design and implementation. This disposes us to a fetish for novelty and a collective blind spot about mundane and old technologies.[8] In the case of LPFM, activists and other social groups reinterpreted radio fully seventy years after its dominant meaning had largely stabilized. The 1990s "micro-broadcasting movement" was a new social group. But the goals of radio activists in the 1990s and 2000s were not entirely distinct from those of broadcast reformers in earlier eras who sought to establish alternatives to networked commercial radio.[9] In keeping with earlier generations of reformers, LPFM's boosters held dissenting viewpoints in their own time period. LPFM advocates' interpretive work has been a significant focus of this book. They provide a unique case within social studies of technology. To a significant degree, they succeeded in redefining the use and meaning small-scale FM broadcasting, as well as reconfiguring radio as an assemblage of analog and digital technologies (used to share and transmit audio over FM spectrum). In so doing, they transfigured an old medium.

My final point is that contestations over the meaning of radio can reveal depth and gradation in the meaning and valuation of electronic communication more generally. Attuning to contemporary controversy over radio helps us triangulate discourses about what we think communication technologies should do. This idea's implications are wider than radio.[10] Listening to a fuller field of objects enables us to pick up on contrast between discourses that may be harder to hear if we train our attention on a particular technology or phenomenon in isolation. Dismissing radio as a "dinosaur" technology because of its supposed eclipse by newer media is unwarranted for a variety of reasons. Radio is, in fact, still undergoing change. Listening to efforts to define an old technology that is distinctly nondigital (symbolically anyway—materially this is a far more complex story because networked computing can be used to extend the sociotechnical arrangement we still call *radio,* and digital signals can transmit satellite and terrestrial radio) can help us hear what issues and values may matter in a wider conversation about communication rights and the media future.

It is worth pointing out that "technology" is neither a moral order nor a material condition that we can take at face value. Statements about technology—celebratory, cautionary, or "merely" descriptive—are rife with claims about social order, power, and values. Learning to listen to how political agendas and values are articulated to artifacts is of paramount importance. The dynamics surrounding radio activists' conscious propagation of specific, prescribed meanings for radio can be observed across myriad forms of technological evangelism (with varying contours, of course). In an era when we are nearly deafened by sloganeering about what "the digital" is or does, a study in which we can carefully attend to how politics come to be "inside" the artifact has significance that extends well beyond radio.

A widely held conception of media technologies is that their purpose or main use is to transmit or communicate "information," "messages," or "content." "Information" is a widely deployed, widely studied, and complex term and concept.[11] The reason that we believe "media technologies" are different from other technologies in the first place has to do with the idea of information.[12] "Information" is a "keyword" in the sense invoked by cultural historian Raymond Williams. Its use to describe and understand media constitutes a significant site for understanding its meaning in the wider culture.[13] Conceptions of media technologies as tools for transmission of information usually encompass both a limited technical sense and the notion that "information" is a social force. Information as a social force relates to the idea that communication technologies are inherently oriented toward progress and moral betterment.[14] Radio, for example, was understood in its early days as a means of fulfilling "social destiny." For many, broadcasting promised the end of demagogy, the advent of a more reflexive polity, and the rise of national unity amid growing diversity.[15]

Statements about information are so ubiquitous in contemporary society that they hardly stand out. It is nonetheless worth taking note of some specific statements about the value of information and the use of technologies to promote and transmit it. This was expressed particularly forcefully by the founder of the community center in Imbaseni, Tanzania, with whom Prometheus built a low power community radio station in 2005 (described in chapter 6). The center's founder, an American former Black Panther living in exile, stated, "I'm particularly impressed with the fact that Prometheus has given [us] this very, very, very powerful tool, this radio station, that will allow us to disseminate information. My old compatriot, Brother Eldridge Cleaver, once said that 'information is the raw material of new ideas.' I'm going to repeat that, *Information is the raw material of new ideas.*"[16] The statement by Representative Maurice Hinchey (D-NY) (quoted in chapter 6) also illustrates the common belief that broadcasting technologies transmit information. Hinchey said that the nation was witnessing the culmination of "a twenty-year right-wing plan to control information people receive…." In both of these statements, information is held to be a social force, and media technology is central to how information is conveyed.

By contrast, the radio activists themselves invoked radio's power to "transmit information" relatively infrequently. They instead privileged the notions that media technologies promoted empowerment, "community," pluralism, and localism. This deviation from the discourse of "information" is significant. It demonstrates skepticism toward a core meaning of electronic communication. In this, the radio activists had perhaps arrived at a similar conclusion to that of philosopher of technology Langdon Winner, who argues that "a serious misconception among computer enthusiasts is the belief that democracy is largely a matter of distributing information…."[17] The radio activists' "information-ambivalent" stance presented an obstacle for them. When they sought to propagate wi-fi in addition to radio, they ran squarely into a potent digital utopian vision they did not fully share.

Drawing out "information" as a value illuminates the policy disputes over webcasting, radio, and other options for electronic communication in the late twentieth and early twenty-first centuries. To reprise a discussion from chapter 6, National Public Radio, a significant player in the debate over the FCC's decision to introduce LPFM, had argued that the main purpose of communication technologies was to transfer "content" between a source of content and a listener, exemplified in this quote: "It makes no practical difference to the listener whether the source of the content is a low-power station transmitting from a mile away or a full service station transmitting from five or ten miles away."[18] NPR went on, "The benefits associated with the [LPFM] proposal may be better realized through other means of electronic communication, such as the Internet. The Internet is revolutionary because it connects individuals with distinct interests, whether those individuals reside within the same town or on opposite sides of the planet...."[19] NPR suggested that what was important in choosing an electronic communication platform or technology was the potential to serve and to cohere communities of interest. They implied that "local" communities bound by geography, ethnos, or other spatiocultural factors were less important. This resulted in NPR proposing that would-be LPFM broadcasters congregate in cyberspace, not the ether. It would be easy to interpret this statement as having no more significance than NPR's desire to deflate and dismiss the goals of low-power radio advocates. Indeed, NPR was doubtless motivated by that agenda.

But there was more at stake in NPR's claims. In privileging "communities of interest" whose members could be distributed anywhere from the same town to the opposite side of the earth, NPR drew on what could be termed an "informational" discourse. Implicitly, this discourse rests on the notion that what is paramount in electronic communication is the exchange of "information." That information is divorced from context, from bodies, from space, and from place.[20] NPR was likely unaware of the full implications of its exhortation to "go on the Internet instead," but this statement reveals why critical attention to the interpretative work surrounding technologies is warranted. This disagreement was about far more than the "purely" technical properties of the respective artifacts. Indeed, efforts to differentiate between the properties of technological artifacts are important because they are never merely descriptive; they serve to bind artifacts to meanings.

The radio activists clearly were uncomfortable with the idea that media technologies are primarily conduits for information. Conversely, other social groups with whom they interacted (such as the community groups and legislators in the previous examples) were more likely to highlight the primacy of "information." In the radio activists' conception, low-power community radio did not simply "transmit information." That framing would leave LPFM open to charges of interchangeability with the Internet. For the activists, "local" or community-level origination of information was an important consideration, which was distinct from the freedom to exchange information in a general sense. They promoted a vision of "community" based on geography and common interest in a locality, as opposed to a geographically dispersed community of interest. Even when the activists drew on the rhetoric of information, they tended to emphasize the significance of community-level creation, transmission, and reception of information as an ideal for electronic communication (a use for which they felt radio was especially apt). The radio activists resisted what they perceived to be unbridled and uncritical enthusiasm for what digital utopianists claimed to be the inherently emancipatory properties of computers and the Internet.

Prometheus's routine promotion of the "community" aspect of community wireless and community radio bears out Raymond Williams's assertion that "community" is imbued with a powerful and positive meaning. It is worth interrogating further what the activists meant by "community," given that the term is utterly vague, save for its positive connotation. Historically, the term marked a contrast between "society" or "the state" and smaller-scale associations (including shared belief, kinship, or shared place).[21] For the radio activists, "community" seemed to combine geographic proximity and shared concerns.

This stands in subtle contrast to the concept as promoted by proponents of networked computing and "virtual community," such as Howard Rheingold and Stewart Brand. They idealized disembodied, geographically distributed networks of users who were nonetheless participants in shared, collaborative, and even intimate sociality.[22] They believed that information wants to be free: free of embodiment, freely flowing, freely commodified and exchanged.[23] The radio activists were more reluctant to "free" their ideals for electronic communication technologies from the local roots and intertwined social networks that might exist in a neighborhood or a municipality. Their commitment to radical politics and time-honored material practices of community organizing such as knocking on doors may have predisposed them to value a notion of bounded localism.[24] The radio activists believed that LPFM was suited to a "community" scale of use, and to the promotion and maintenance of that community. For them, LPFM and "community" were mutually reinforcing. (It is interesting to note that radio activists and digital utopianists share a common heritage in Appropriate Technology, but took its notions about the transformative potential for relationships with technologies in rather different directions.) In essence, even as the radio activists recognized the ability for communications networks to be distributed freely in time and space, they remained committed to the notion of proximal community, with its immediacy, contextual meaning, and relationships between neighboring bodies (bodies in all senses).

The point here is not that the radio activists are "right" and that digital utopianists are "wrong." It is that careful attention to discursive claims around the meaning and value of different technologies should be a goal of scholars, policy advocates, and propagators of technology. We should make every effort to understand assertions about the properties of a given technology for electronic communication as rhetorical claims. They are claims about values as much they are as technical descriptions. That said, the radio activists' attempt to vest debates about electronic communication with a value of proximate and bounded sociality may capture something missing from much strong digital utopianist rhetoric and something we may lose if we concentrate too greatly on listening to (and building for) information utopianism.

Throughout the book, I have largely left underspecified the radio activists' ultimate goals. Although they were not proponents of digital utopianism, they certainly held romantic notions about the capacity for democratic social relations, participatory politics, and sweeping progressive change. They hoped that these changes would be direct results of robust community media infrastructure and participatory expertise (though again, the specific contours of these imagined social relations were rarely articulated very explicitly). As discussed in chapter 6, the radio activists routinely glossed over the issue of whether they wanted to propagate radio because of its potential to promote certain forms of content, or whether they felt there were inherently emancipatory properties to LPFM. While the radio activists were not *digital* utopianists, they could still be characterized as *techno*-utopianists. Their information ambivalence is consonant with this assessment; rather than focusing on radio as a conduit for "content," they focused on the moral potential of the technology more broadly.

In focusing on the utopian promise of LPFM, the radio activists left the potential for regressive uses of radio steadfastly unexamined. In some respects, this may have been a conscious and strategic decision. Deeply examining the agendas of right-wing allies in the legislative struggle to expand LPFM could have undermined Prometheus's ability to partner with a politically diverse range of champions for LPFM. Yet there is certainly more to it than this. I observed one telling exchange in which a visitor to Prometheus's office invoked the putative role of radio stations in the Rwandan genocide in 1994. Activists did not challenge her outright, but stated instead that "it wasn't just the radio stations, but all the media" advocating violence.[25] Empirically, there is validity to this claim.[26] But it stood out because it was rare for the radio activists to consciously consider, let alone acknowledge, a potentially undemocratic or regressive outcome of their work. The possibility that "community governance" might result in parochialism or even genocide had little place in their understanding of small-scale broadcasting. This is a limitation of thinking like an evangelist too much of the time. Utopian and dystopian rhetorics both lack nuance and sophistication in their abilities to capture the contours and consequences of a given sociotechnical arrangement.[27] Instead, it is precisely careful analysis and thoughtful consideration that we most need in order to puzzle through these incredibly thorny questions of technical and political change.

The reader may have noticed that several of the chapters in this book opened with accounts of failure or conflict. These narrative devices are not intended to denigrate the radio activists' efforts. In part, they are meant to illustrate that the work of social change is hard, halting, and slow. Our social and material worlds require production, maintenance, and repair. In real time, material and social metamorphoses can feel elusive. That being said, there are occasional lurches of motion. On November 30, 2012, as I was writing the conclusion of this book, the FCC formally adopted new rules for LPFM. These rules named the technical specifications that would shape the new round of licenses. They would allow stations to be closer together on the FM dial. The radio activists had won their regulatory battle and were poised to begin a new round of licensing and station building.

We can take failures and conflicts as well as successes as points of entry into other conversations we should be having about technology in culture. Even if the radio activists did not repair the big transmitters, what did they produce instead? Fundamentally, they were engaged in cultural mediation of technology, which potentially had profound effects on how everyday people might understand, approach, and use technology. These acts of mediation surround us, so much so that we can fail to notice them. An ethnographic account of media activism offers color and specificity within a landscape littered with overgeneralizations and hyperbole about the impact of media technologies (and especially, the newest media technologies). The story of West Philadelphia radical "weird-beards" going into the wilds of Washington, DC, and pastoral Tennessee, soldering irons aloft, is highly idiosyncratic. Yet the point was never to be anything but historically and anthropologically particular (at times even peculiar). At the same time, the weird-beards' story helps us to recognize that there are politics embedded in our artifacts. It also shows us how our political beliefs draw us to these artifacts. In this recognition, debate and transformation of both become possible.

Notes

Introduction

1. Juris (2008), 17.

2. "Propagators" represents a refinement of the category "mediators" of technology. See Goldstein (1997); Greenberg (2008); Kline (2000).

3. They are also "broadcasters" in the sense of widely spreading or disseminating radio technology.

4. See Hess (2005) for an account of what he calls "technological- or product-oriented movements" (TPMs).

5. See Coleman (2012); DeNardis (2009); Kelty (2008).

6. Negroponte (1995), 230. Scholars, too, have advanced such claims: Yochai Benkler writes that "the networked information environment offers us a more attractive cultural production system in two distinct ways: (1) it makes culture more transparent, and (2) it makes culture more malleable" (2006), 15; see also Castells (1996); Lessig (1999). Thanks to Lucas Graves and Tom Streeter for comments and conversations about these issues.

7. Thanks to an anonymous reviewer for this observation.

8. Juris (2008); Wolfson (2014).

9. McChesney (2004).

10. Haring (2006); Douglas (1987); Blake (2011).

11. Pursell (1993); Turner (2006); see also Dunbar-Hester (2008).

12. Horwitz (1997); Pickard (2011).

13. See Gieryn (1983).

14. Of course here I am referencing Langdon Winner's famous essay, "Do Artifacts Have Politics?" With gratitude to Mike Lynch for the inversion.

15. Gusterson (1996).

16. Pooley (2011), 230, footnote 14. See also Napoli (2009), 23–25. Assuming it is a movement, it is alternately called "media reform movement," "media democracy movement," or "media justice movement," depending on which movement actors one asks; unsurprisingly, the label around which groups orient matters greatly to them and marks position in a variety of ways.

17. Woodhouse et al. (2002), 298–299.

18. Carroll and Hackett (2006), 88.

19. Melucci (1996); Carroll and Hackett (2006).

20. Carroll and Hackett (2006), 93.

21. Polletta (2006); see also Juris (2008). Conceptual tools developed by scholars of social movements such as political opportunity structures, resource mobilization, and frames of meaning are essentially orthogonal to the dynamics that I draw out (McAdam, Tarrow, and Tilly 2001; Tarrow 1998).

22. Bijker (1994), 77. Rayvon Fouché (2006) has pointed out some limitations of tracing networks of groups because of the fact that marginalized or oppressed groups may not be included in these networks.

23. Bijker (1994, 78) further states that the "relevant social group" is both an actor's and an analyst's category, and that when following the actor's "identifications, definitions, and delineations, it is the actors' relevant social groups we get" but that these groups are also relevant for the analyst.

24. Clifford (1986).

25. Scott, Richards, and Martin (1990).

26. Gusterson (1996), 13.

27. Clifford (1986), 6.

28. Traweek (1988), 10.

29. Collins and Evans (2002), 254.

30. Ibid.

31. See Boyer (2003); Hannerz (1998); Boyer and Hannerz (2006).

32. Informal conversation (November 2007).

33. E-mail, Jasper to author (February 16, 2006).

34. Juris (2008), 20.

35. In earlier presentations of this research, I referred to the radio activists' organization as Pandora Radio Project, a thinly veiled version of their actual name. This was more my (and Human Subjects Review) preference than the activists'; in our discussions of my research over the years, they were largely indifferent to whether their organization was named in public presentations of my research. One of them actually generated Pandora as a pseudonym, but said he could go either way over whether I used it, their real name, or something else. (Incidentally, my use of this pseudonym in presentations and papers precedes the appearance and ascent of the online music service Pandora Internet Radio. The sites are in no way related or affiliated.) One other small advantage to naming the organization is that I am no longer saddled with the distortion of Greek mythology that "Pandora" had caused. Prometheus, of course, stole heavenly fire from the gods (or commercial broadcast industry, or the FCC). (The activists were occasionally amused by the idea that they might have unintentionally unleashed all the evils of the world, as suggested by "Pandora," but it is preferable to accurately depict the symbolism of their name.)

36. Anonymity does not always serve the interests of researchers or their subjects of study (Stein 2010). My thanks to Dan Kreiss, Victor Pickard, and Todd Wolfson for conversations about this topic.

Chapter 1

1. Deleuze and Guattari (1987). IMCs are loosely affiliated but not hierarchically organized or maintained.

2. The Indymedia critique of neoliberal globalization emanated from the Zapatista resistance movement in Mexico in 1996. Todd Wolfson (2014) writes, "Heralding the coming of a new global politics of resistance, the Zapatista Army of National Liberation (EZLN) made a widespread call to activists, revolutionaries and media makers to forge 'an intercontinental network of struggle against neoliberalism' and 'a network of *communication* among all our struggles'" (emphasis in original).

3. Wolfson (2014). John Downing (2003) estimated that there were about ninety IMCs worldwide in 2002.

4. Carroll and Hackett (2006), 88.

5. Ruggiero (1999), 18.

6. Klinenberg (2007), 62.

7. Of course *citizen* is rightly a contentious concept for some. In using this term occasionally, I wish to signal activity around civic and communal participation, not to marginalize those without full legal status as citizens. This is important because many media activists possess a wider social justice orientation, including immigra-

tion rights. Several of the LPFMs Prometheus built were with migrant farmworker groups, as well. I do not have space to interrogate "citizenship" here, but using it to stand in for a mode of engagement open to "everyone" may present problems.

8. See Coopman (1999) and Shields and Ogles (1995) for histories of these legal challenges, notably those mounted by Mbanna Kantako and Stephen Dunifer.

9. Piette (1998); Karr (2005).

10. See Hazen and Winokur (1997); McChesney et al. (2005).

11. Carroll and Hackett (2006), 86.

12. Interview (July 2003).

13. Cyril (May 13, 2005).

14. Douglas (1999): 219, conclusion.

15. REC Networks, "LPFM Scoreboard," http://cdbs.recnet.net:8080/scoreboard.php (January 12, 2012).

16. See Dick and McDowell (2000): 331.

17. This is complex: LPFM stations are not required to broadcast local content, but in cases in which more than one group is in competition for a license, locally originating content is a factor in the FCC's decision regarding which group will receive the license.

18. Some legal LPFMs have adopted the "free radio" mantle when choosing their names, marking continuity with the movement for microradio in the 1980s and 1990s, as well as the goals of making radio free and accessible. And the term *free radio,* especially when used in conjunction with a place name, such as *Free Radio Santa Cruz,* also seems to carry a linguistic reference to, and inversion of, the US-funded Cold War propaganda service in Europe, Radio Free Europe. Finally, some have suggested that the origin of the term *free radio* is from the 1960s US counterculture, invoking, for example, *free love* (*Wikipedia* April 8, 2011). Adding credence to this notion, Abbie Hoffman's 1970 countercultural classic *Steal This Book* contains a section on radio in the chapter "Free Communication."

19. In the United States, radio spectrum is understood as a publicly owned resource that is licensed by the FCC, a federal regulatory body established in 1934 (after its introduction as the Federal Radio Commission in 1927).

20. Slotten (2000), 117–118.

21. Ibid., chapter 4.

22. Riismandel (2002), 429.

23. Riismandel (2002), 429. According to Riismandel, between 1949 and 1956, the FCC granted 245 commercial FM licenses but deleted 722.

24. Horwitz (1989), 177.

25. Riismandel (2002), 429.

26. Keith (2002), 393.

27. Slotten (2000), 113.

28. Horwitz (1997); Pickard (2011).

29. Horwitz (1997), 311–312. See also Boyle (1997); Downing (1984, 2000); Halleck (2002); Kidd (2002 a) on precursors to the 1990s radical media movements.

30. Public Broadcasting Act of 1967, Subpart D—Corporation for Public Broadcasting Sec. 396. [47 U.S.C. 396] Corporation for Public Broadcasting.

31. Horwitz (1989), 251. See also Boyle (1997) for an account of what she calls "guerilla television," video collectives working in the 1960s to 1970s.

32. Riismandel (2002), 430.

33. Riismandel (2002), 431–432.

34. In 1994, the FCC began to hold auctions to allocate frequencies to eligible parties, which literally raised the cost of licenses. For most of the FCC's history, licenses were allocated through comparative hearings; yet licenses could be transferred or sold by license holders. Starting in 1982, some allocations were made by choosing among eligible applicants by random lottery (Gattuso 1989). Even before the cost of licenses rose through the auctioning system, some amount of revenue was required to obtain a license, either to hire lawyers and engineers to prepare documents for hearings or to purchase a license from its current holder. For all these reasons, obtaining even noncommercial licenses became prohibitively expensive for the types of groups who would have prior to 1978 been eligible for Class D licenses.

35. Walker (2001), 214. See also Brand (2004); Coopman (1995): chapter 2; Soley (1998).

36. Duncan (1998).

37. Janssen (2003); RECNET Scorecard, http://cdbs.recnet.net:8080/scoreboard.php (April 8, 2011).

38. Brecht (1993).

39. Douglas (1999), 330. Emphasis in original.

40. Haring (2006), 29. Though the denial of politics was a dominant motif in ham culture, it was not a universal value; in 1961, a ham named Fred Huntley drew ire from the ham community at large when he started an anticommunist amateur network. (His fellow hams did not claim to *endorse* communism, naturally, but advocated other forms of "neutral" and "appropriate" communication [Ibid., 108–109].)

41. Douglas (1987), chapter 6.

42. Ibid.; Haring (2006).

43. Haring (2006), 156.

44. Blake (2011).

45. Ibid., 544.

46. Haring (2006), 156.

47. Ibid.

48. Pursell (1993), 636.

49. Ibid., 634.

50. Lorenzo W. Milam, *Sex and Broadcasting: A Handbook on Starting a Radio Station for the Community* (Saratoga, CA: Dildo Press, 1975).

51. Prometheus Radio Project website, http://www.Prometheusradio.org/node/2006 (July 5, 2013).

52. Hoffman (1996/1970).

53. Winner (1988).

54. Field notes; see also Klinenberg (2007), 249–250.

55. Quoted in Karr (2005): 25.

56. Steinberg (2003).

57. E-mail to basement (April 15, 2005).

58. Klinenberg (2007), 62.

59. McChesney et al. (2005), 367.

60. Coyer and Tridish (2005).

61. The 2004 brouhaha over indecency prompted by the exposure of Janet Jackson's breast during the Super Bowl represented another instance of public concern over broadcast policies. And some blamed the lowering of decency standards on broadcast media on corporate consolidation: Representative Edward J. Markey, D-MA, stated, "There's no real deterrent effect left. This is especially true of the multibillion conglomerates that control a multitude of stations." CBS News (February 11, 2004). See also Klinenberg (2007), 207–208.

62. National Public Radio (March 30, 2000). See also Klinenberg (2007), 257.

63. See Janssen (2001a; 2001b). LPFMs were already required to maintain the more stringent (third-adjacent channel) protection from radio signals transmitting reading services for visually impaired people.

64. Walker (2001), 260–264.

65. FCC (March 24, 2000).

66. Janssen (2001a; 2001b).

67. MITRE Corporation (2003).

68. Janssen (2003).

69. Riismandel (2002), 441–442.

70. Ibid., 442.

71. United States Senate, 109th Congress, 1st Session (February 8, 2005). Emphasis added. The Senate did not pass the bill. All along, Senator McCain was very critical of the "back-door" means of using the appropriations bill to pass this legislation, stating that at the time "there is no way they could have carried that vote on the floor of this Senate. There is no way they could have deprived all of these communities, all of these small business people, all of these religious organizations, all of these minority groups—but they stuck it into an appropriations bill, a piece of legislation that never had a single bit of debate and would never have passed through the Commerce Committee, of which I am the chairman, if it had been put to a vote" (Senator John McCain, remarks on the floor of the Senate, October 26, 2000).

72. National Association of Broadcasters (October 13, 2003): 12.

73. I do not wish to imply that political and technical domains are ever truly separate.

74. The FCC commissioners change fairly regularly. Commissioners are nominated for a five-year term by the president, who also designates one to serve as chairman. Only three may be of the same political party. This tends to mean that the commission is usually made up of three members, including the chairman, who are of the same political party as the president. The backdrop in the LPFM story is the shift in political power from the Clinton administration with a Republican majority in Congress, to the G. W. Bush administration with a Republican majority, until 2006 when the Democrats narrowly gained a majority in both houses of Congress. Attendantly, the chairman under Clinton (William Kennard) was a Democratic appointee and the chairmen under Bush (Michael Powell, Kevin Martin) were Republican appointees. Many people I interviewed in Washington spoke of the LPFM conflict between Congress and the FCC as having to do with partisan opposition to the Clinton administration.

75. See Latour and Woolgar (1979).

76. National Association of Broadcasters (September 2004).

77. RECNET, "LPFM Scoreboard," http://recnet.net/scoreboard.php?wind=1; Janssen (2003).

78. The astute reader will note that this slogan closely resembles the title of this book. I borrowed this phrase from Radio Free Nashville, as did Eric Klinenberg for a chapter in his 2007 book *Fighting for Air*, but it is something of a meme in the low-power radio world, and original attribution is all but impossible. (See also Ruggiero 1999.)

Chapter 2

1. See Woolgar (1991); Akrich (1995); Kline and Pinch (1996); Oudshoorn and Pinch (2003); Fouché (2006).

2. See Cowan (1987); Hughes (1987).

3. See Goldstein (1997); Greenberg (2008); Kline (2000).

4. I say "technological engagement" rather than "technical skill" because other mediators—such as video retailers in Greenberg's (2008) example and the home economists discussed by Goldstein (1997)—also would have been available to teach consumers how to operate equipment. It is worth noting that I use "technical" mainly as a descriptive category based on its salience for the actors, for whom "the technical" was of great symbolic importance. I do not mean to suggest that many of the actors' practices or artifacts *not* associated with electronics, radio hardware, or software are not "technical" in a broad sense, but this is not an analytical distinction worth making here.

5. Quoted in Karr (2005), 24. This also explains why the radio activists did not particularly attend to notions of listeners or audience, who could certainly also be theorized as "users'" or "consumers" of LPFM; they were more concerned with the idea of citizens and "ordinary people" "having a voice." Listeners received relatively scant attention from the activists, which is mirrored in this analysis.

6. See Hess (2005).

7. Butler (1990), 142.

8. Butler (1993).

9. However, with a category such as gender or race, the classification is constructed through not only identity but also social structures, institutions, and representation. This chapter focuses on identity work, performances of identity that are elective, due to individual agency, as well as more unconscious or seemingly unconscious

displays, at the expense of more structural factors that may work to impose identity or consciousness of group membership on individuals (Haraway 1991, 155). This is not because structural factors are unimportant but because this analysis highlights the actors' attempts to enact certain values through relationships with technology, while pointing to the potential limitations of or complications raised by using identity work (alone) to change identifications such as race or gender.

10. Gergen (1991); Turkle (1995); Haring (2006).

11. See Orr (1990, 1996); Douglas (1987); Turkle (1995); Haring (2006).

12. My discussion of the radio activists' identity performances is also indebted to sociologist Erving Goffman's (1959) and anthropologist Victor Turner's (1987) work on the performative aspects of communication and interaction.

13. Field notes (August 18, 2006).

14. Turner (2006), 4, passim. Turner notes that the New Communalists themselves owed a debt to romantic and transcendentalist ideals (62, 75). Carroll Pursell states that "the culture of Appropriate Technology ... was more than a little reminiscent of two constructions of masculinity which were widely adhered to in the United States at the beginning of the 19th century:... the republican gentleman ... and the independent producer" (1993, 636).

15. New Communalists favored cooperative living arrangements in urban and rural environments (Turner 2006, 74).

16. Field notes (March 2005).

17. The activists were generally not adherents to frequent showering, an antibourgeois identity display.

18. Field notes (July 20, 2005).

19. Beards and body hair are, of course, a means of performing gender as well.

20. See Gelber (1997) for a discussion of the historical origins of DIY and Waksman (2004) for the intersection of technological enthusiasm and DIY among electric guitar musicians. See Portwood-Stacer (2013) for extensive consideration of radical anarchist lifestyle, identity, and self-presentation.

21. Interview (July 5, 2006).

22. Indeed, the unkempt beard was sported not only by back-to-the-landers but also aesthetically and sociologically overlapping technological counterculture "grandfathers" including free software's Richard Stallman.

23. Field notes (September, 18, 2004).

24. Field notes (February 24, 2006).

25. Field notes (November 3, 2004).

26. Their website also provides links to other "sustainable energy radio stations," which is important because it not only aligns them with Appropriate Technology values but gestures towards a network of similar stations, as opposed to viewing themselves as a lone hydropowered station. http://www.wjffradio.org/wjff/index.php?section=19 (March 11, 2008).

27. Pursell 1993, 632.

28. Interview (June 27, 2006).

29. Turner (2006), 71.

30. *Journal of Broadcasting* 15, no. 4 (1971), review of *Sex and Broadcasting,* quoted on back cover of Milam 1975.

31. Turner (2006), 73.

32. Milam quoted in Cruickshank (2002). Prometheus staff also claimed that when *Sex and Broadcasting* went out of print, Milam gave them several cases of the last available copies, which they offered for sale on their website.

33. KPFA is the first station of the Pacifica network, founded in 1949 in San Francisco by pacifist Lewis Hill (Lasar 2000). KRAB was the first station in what came to be known as the KRAB nebula, loosely affiliated stations including KTAO (Taos, New Mexico, mentioned previously) and KBOO (Portland, Oregon).

34. Walker (2001): 69–70. Emphasis in original. Walker's quote from Milam is in fact from an edition of the *Whole Earth Review* from 1990.

35. Interview (June 29, 2006). See chapter 7 for more attention to discourses of "information" and comparisons between the radio activists' work on LPFM, community wireless, and spectrum reform.

36. Turner (2006), 133.

37. For a 1987 edition (published with Nelson's later *Dream Machines*), Stewart Brand wrote the foreword.

38. Thomas Streeter notes that the modifier "personal" coexisted with micro-, home, or desktop computer, and argues that "personal computer" won out because it was the opposite of "impersonal" computer (2011, 63). He points out that we do not use this appellation for other intimate, personal technologies; for example, we do not refer to wristwatches as "personal clocks."

39. Interview (June 29, 2006).

40. Turner rightly distinguishes between the New Communalists and the New Left. He argues that the strategies for each group differed: "With the New Left, true community and the end of alienation were usually though to be the result of political

activity, rather than a form of politics in their own right [as they were for the New Communalists]" (2006, 35). It is essential to not conflate activism and activist identity with countercultural identity and heritage.

41. Autonomous Food Not Bombs chapters in different cities provide free vegetarian and vegan food and fight war and poverty.

42. Field notes (April 2, 2005).

43. Similar to the *Whole Earth Catalog,* a barnraising is a forum "where a geographically dispersed collection of individuals and groups could come together ... and recognize each other as members of a single community (Turner 2006, 89). The radio activists, however, placed an emphasis on the "real-world" aspect of community as opposed to virtual community.

44. *Oxford English Dictionary,* new edition. Online.

45. This resonates with other iterations of identity politics, such as "queer." Judith Butler points to a tension for these terms of exclusion; even as they are reclaimed and vested with a "positive resignification" (1993, 223), a total metamorphosis, in which past derogation is cast off, may serve to vitiate their full significance. She cautions that "normalizing the queer would be, after all, its sad finish" (1994: 21).

46. Eglash (2002): footnote 1.

47. The *OED* notes that *nerd* has also acquired a definition as a person who pursues a "highly technical interest with obsessive or exclusive dedication." However, it is still more likely to be depreciative, and it is also more broadly defined as "an insignificant, foolish, or socially inept person; a person who is boringly conventional or studious."

48. Eglash (2002); Dunbar-Hester (2008).

49. Haring (2006).

50. Ibid., chapter 6.

51. Thanks to Ronald Kline for pointing out that Dick is another textual link to the 1960s.

52. See Oldenziel (2004): chapter 5; Douglas (1987).

53. And not only Radio Boys: One geek went to great lengths to find on the Internet a "Radio Girls" book for his girlfriend's birthday. The Radio Girls book was from the same series as the Radio Boys, but the girls' books are rarer.

54. Turkle (1984); Håpnes and Sørenson (1995); see also Levy (1984); Himanen (2001). In the early 1980s, Sherry Turkle addressed the relationships hackers form with computers: "computers have become more than a job or an object of study, they have become a way of life" (Turkle 1984, 200).

55. Turkle (1984), 232.

56. Ibid., 235.

57. Gabriella Coleman (2004) has explored the politics of free and open source software (FOSS) developers. She argues that, similar to hackers, FOSS developers' politics are rather more inchoate, agnostic, and informal, as well as self-consciously divorced by the developers from their technical practice. This differentiates them significantly from the radio activists who were deliberate and reflective about their politics and the political implications of their engagement in technical practices. There is, of course, a diversity of hacker politics, exemplified in hackers' efforts around projects ranging from organized crime and corporate espionage to electronic civil disobedience (distributed denial of service, or DDoS attacks) against corporations who hindered the radical journalistic group Wikileaks in 2010, to providing support for the anticorporatist Occupy Wall Street movement in 2011.

58. Interview (July 2003).

59. Interview (July 10, 2003). This brings to mind Carolyn Marvin's work on "inventing the expert," in which experts define themselves in opposition to an ignorant public (1988, chapter 1).

60. It should be emphasized that these politics are not shared by all geeks; many geeks' attitudes about expertise and participation are significantly different from these radio geeks'. When I use the term *geeks* or *the geeks* I am referring only to the local case of these particular radio geeks.

61. Interview September 26, 2006.

62. Himanen (2001), 3.

63. See Kleif and Faulkner (2003); Lindsay (2003), 46. See also Oldenziel (1997) for a discussion of the making of a male technical domain through leisure projects for adolescent boys.

64. John Dougherty (2005).

65. Workshop schedule, Woodburn, Oregon (August 18–20, 2006).

66. E-mail, Renée to basement (September 26, 2006). Italics added.

67. Interview (July 5, 2006).

68. Ibid.

69. Ibid.

70. See Wynne (1996) and Epstein (1997) for discussions of the interactions between lay and expert knowledge.

71. Abbie Hoffman's countercultural classic *Steal This Book* contains a section on radio in the chapter on "Free Communications." Thanks to Peter Sachs Collopy for bringing this to my attention.

72. Interview (July 2003).

73. Interview (July 2003).

74. Boyer (2003).

75. Notably, the geeks' DIY ethic here applied not only to the traditionally masculine realm of FM tinkering but also to the traditionally feminine one of childbirth and midwifery. And, of course, the act of scavenging for equipment is also notable for its display of the values of reuse and self-sufficiency, as well as a technical competence that enabled the actors to discern which pieces of equipment that could actually be useful to them.

76. Interview (July 5, 2006).

77. Prometheus only assisted legal efforts in the United States. Internationally they were much more willing to work in gray areas. One activist said, "We reserve our pirating for other countries" (Field notes March 9, 2005).

78. The FCC's original language prohibited from obtaining licenses only former unlicensed broadcasters who had not ceased broadcasting illegally when requested to do so or by a certain date. The Radio Broadcasting Act of 2000, for which the NAB lobbied, "prohibited anyone who had ever broadcast without a license from applying for a low power radio station" (Microradio Implementation Project website); however the US Court of Appeals for the DC circuit overturned this broad ban in 2002, and upheld the ban only against people who did not cease to operate their illegal stations when told to do so by the FCC. (See LPFM application, online at MonsterFM Broadcast Technical Services, http://monsterfm.com/lpfm/lpfm-application.htm.) By contrast, in the United Kingdom, former pirates have not only been consulted in drafting the UK community radio service (similar to LPFM) but also serve as regulating officials. Lawrie Hallett, the senior associate of the Radio Planning and Licensing Team for the Office of Communications (OFCOM), had in his earlier days been a pirate broadcaster. Sarah Champion quotes Hallett: "I'm now on the other side of the fence—poacher turned gamekeeper if you like" (Champion 2005).

79. Interview (June 23, 2003). The court cases of Mbanna Kantako and Stephen Dunifer engage these issues explicitly. See Shields and Ogles (1995); Coopman (1999); Riismandel (2002); Anderson (2004).

80. Interview (July 5, 2006).

81. The FCC gives preference in licensing to community groups that have been in existence prior to applying for a license, though it assiduously avoids passing judgment on which types of community groups are more worthy than others.

82. Workshop schedule, Woodburn, Oregon (August 18–20, 2006).

83. Field notes (August 19, 2006).

84. Turner (1969).

85. Naturally, the radio activists turned to radio to make this happen. For workshops and the plenary addresses, they set up a microphone, transmitter, and headset receivers so each presenter could speak in her native language and have her speech translated in real time and broadcast by bilingual volunteers. Audience members who were not bilingual could tune into the event in their own language and listen to translations on a headset receiver.

86. Press release, Prometheus Radio Project (March 28, 2005).

87. Interview (July 5, 2006).

88. See also Keniston (1968) for discussion of distinctions between "committed" radicalism versus lifestyle or countercultural forms of involvement.

89. In a board meeting, Jasper said, "LPFM isn't a wonk issue; it's something people go to jail over!" Field notes (February 6, 2005).

90. Field notes (March 16, 2005). See Portwood-Stacer 2013.

91. We might think of radio here as a "liminal entity": "Liminal entities are 'neither here nor there; they are betwixt and between positions assigned and arrayed by law, custom, convention, and ceremony'" (Victor Turner quoted in Pinch and Trocco (2002), 308).

Chapter 3

1. Interview (July 25, 2006).

2. My choice to consider gender in a separate chapter should not be interpreted as a belief that gender is produced in isolation from other categories of identity such as those explored in chapter 2. Rather, the evidence here is abundant and knotty enough that it warrants consideration as its own topic. Thus this chapter performs a delicate (and partially reluctant) act of isolating gender in order to hear the ways it was produced and the work it did in the context of radio activism. Chapter 2's insights into the entanglement of activist, countercultural, and geeky aspects of identities produced in relation to radio technology can still be heard, though here they are muted for the sake of listening to gender; some signals are tuned out in order to attend better to one.

3. Oldenziel (2004); Kleif and Faulkner (2003).

4. It is not my intent to overlook the importance of structural or institutional factors or the importance of representation in culture as a means of perpetuating them (Lerman, Mohun, and Oldenziel 2003). The focus here, rather, is the role identity work plays in (re)imagining a particular role for radio technology and technical skills. (Additionally, though this analysis highlights identity work, the activists'

goals were not restricted to this register: their use of identity work to reinterpret radio technology, including tinkering with gender and promoting geekiness, was intended to influence social structure as well.) The reason for this treatment of gender is twofold: (1) this book attends to a wider array of identity performances, and this treatment of gender allows it to be considered in relation to the other aspects of identity explored in the preceding chapter; (2) the barnraising site and Geek Group site together constitute a wealth of empirical material with regard to gender identity work, and the activists' hoped-for as well as actual gender constructions present a fruitful and provocative locus to analyze the intersection of gender and technical practice.

5. Wajcman (1991).

6. Kline and Pinch (1996), 780; Lerman, Mohun, and Oldenziel (2003), 4.

7. Douglas (1999), 16–17. This should not be taken to imply that there is a *single* technical masculinity, either over time or at any particular historical moment. See Wajcman (1991), chapter 6; Horowitz (2001).

8. Activists focused on decoupling both technical skill sets and domestic skill sets from gender identity, though they focused more on the technical. I do not mean to suggest that domestic work is divorced from technology or technical skill. This is partly an actors' label; the Prometheus group used *technical* to refer to audio, computer, and radio transmission hardware and software. Unless otherwise specified, *technical* here means related to radio hardware and equipment.

9. Indeed, the men in the group generally identified as feminist men who openly sought to challenge masculine hegemony. See Connell (2005); Digby (1998).

10. Age could have also been a factor here: the electrician was probably in his fifties, and the person he was looking for, like me, was in her late twenties. I have less means to compare class or rural-metropolitan factors in this instance, though the electrician was local to the Nashville area and Louisa and I had traveled from Philadelphia.

11. Men and women used this sort of language to describe technical or, interestingly, activist prowess.

12. Interview (September 26, 2006).

13. Interview (July 2003).

14. Indeed, sociologist Erving Goffman (1959) might call the barnraising a "front stage" event overall.

15. Richard Schechner quoted by Victor Turner (1987), 76.

16. E-mail, Brian to basement (November 1, 2005).

17. Interview (July 5, 2006).

18. Ibid.

19. Ibid.

20. I speculate that "dudecore" comes from "hardcore," substituting the "hard" with another signifier for masculinity (see Edwards 1990). It also sounds like "corps," as in "Marine Corps." It did not have an entry in the *OED, Wikipedia,* or the online Urban Slang dictionary when I checked. Among these actors, the term *dude* was commonly used to refer to men, as one can note in the quotes.

21. Dougherty (2005).

22. Ibid., 5.

23. Interview (July 27, 2006).

24. Interview (July 25, 2006).

25. Ibid.

26. Dougherty (2005).

27. Interview (July 25, 2006).

28. Interview (June 27, 2006).

29. Turner (2006), 77. Italics in original.

30. Interview (July 25, 2006).

31. Pinch and Trocco (2002, 138) speculate that this may have been the case with synthesizer player Wendy (formerly Walter) Carlos. Sherry Turkle has also discussed mutable performances of gender in relation to technology, but the radio activists present a significantly different case because their gender performances occur in nonvirtual spaces (1995, chapter 8).

32. Interview (June 27, 2006).

33. Ibid.

34. Pursell (1993): 636. He adds that the dominant culture interpreted Appropriate Technology not as a masculine undertaking, but rather as a feminized attitude toward technology, arguing that Appropriate Technology may have suffered a backlash by a wider culture that wished to embrace technological vanguardism, including high-tech weaponry, as a symbol of national virility and power (ibid.). This points to the complexity of the gender-technology relation as it intersects with alternative politics and countercultural lifestyles.

35. See Kline and Pinch (1996).

36. It would be foolish to discount the roles played by structures and institutions in constraining (or promoting) equality in terms of race or gender; consideration of

identity alone would be critically lacking as a single explanatory mechanism for understanding the inequities in achievement of technical skills for women, people of color, or people in cultural contexts besides the Global North, of course. The community groups that Prometheus assisted were not restricted to mainly white groups.

Chapter 4

1. For accounts of media activism, see Carroll and Hackett (2006); Downing (2000, 2003); Mueller (2002); Pickard (2008); McLaughlin and Pickard (2005). See Hess (2005) for an overview of technological activism and Hintz and Milan (2010) for a description of grassroots technological media activism. Unlike this account, these are more meso- and macro- level analyses that address typologies of media activism, not practice.

2. Barley and Kunda (2001), 81.

3. Orr (1996), 10.

4. This account is also indebted to accounts of science practice that closely attend craft, practice, rituals, tacit assumptions, and rhetorical forms in order to contextualize and deconstruct the scientists' "final products" and narratives of achievement, certainty, and consensus. Classic accounts include Latour and Woolgar (1979) and Knorr Cetina (1999). For analyses detailing work with machines as technical practice, see Orlikowski (2000); Orr (1996); Suchman (1987).

5. *Prometheus DeLivered* (2005), 9.

6. Ibid.

7. Field notes (May 28, 2005).

8. Ibid.

9. Ibid.

10. Field notes (May 29, 2005).

11. Casper (1997): 235.

12. For example, Ellen to basement (January 19, 2006).

13. Interview (July 5, 2006). In his reference to "old-school dude engineers," Brian revealed awareness of gender dynamics that made inclusion of novices, especially women, more challenging.

14. Interview (February 16, 2006).

15. See Carroll and Hackett (2006); Juris (2008).

16. Field notes (August 18, 2006).

17. Retrieved from Prometheus website, http://www.Prometheusradio.org/communityradio (October 11, 2012). *Community* is of course notoriously difficult to define. Raymond Williams notes that *community* is a curious concept in that it does not take on a negative cast: it may "be the warmly persuasive term to describe an existing set of relationships, or the warmly persuasive term to describe an alternative set of relationships" (1976, 76).

18. Kelty (2005), 200.

19. See Postigo (2003); Terranova (2004); Turner (2009); Scholz (2012). Although this has been prominently documented in "the digital," it is not solely a phenomenon of "the digital."

20. Haring (2006): 93; see also Douglas (1987), chapter six.

21. Williams (1976): 334–335; Orr (1996), 9–10.

22. See Rosenzweig (2006) on amateur production in the context of *Wikipedia*. See also Jenkins (2006).

23. See Antonijevic, Dormans, and Wyatt (2010) for a discussion of feeling and affective labor in academic work.

24. Interview (February 16, 2006).

25. Orr (1996), 10.

26. See Douglas (1987); Dunbar-Hester (2008); Kleif and Faulkner (2003).

27. Kreiss, Finn, and Turner (2010): 244.

28. Turner has harsh words for a similar dynamic in the New Communalists' case. He argues that Ken Kesey, Stewart Brand, and others acted as alpha males while downplaying the power they wielded (2006, 90). Without explicit efforts to combat such dynamics, it is very hard for activists and others committed to nonhierarchical collaboration to prevent such dynamics from creeping in. Of course, Turner's point is that for the New Communalist alphas, the commitment to nonhierarchical organization was only nominal.

29. Coleman (2004), footnote 10.

30. Independent Media Centers, Anonymous, Riseup are activist technical projects that struggle with the politics of inclusion-exclusion of people with differing levels of expertise, whereas Tor and open source software projects have tended to be composed more uniformly of technical experts, thus obviating some of the conflicts between engineers and technical laypeople. In addition, there is a growing open source hardware movement, which may bear closer comparison to technological media activism than software projects.

31. Jenkins (2006). Fish, Murillo, Nguyen, Panofsky, and Kelty (2011) make the useful point that it may be unproductive to generalize about peer production.

32. See, respectively, Downing (2003); Walker (2001), chapter 8; Gelber (1997); and Waksman (2004).

33. Haring (2006), 8.

34. Haring writes of ham radio clubs: "Clubs grounded hobbyist values in a visible social unit and provided vital mechanisms for enculturation" (2006: 33).

35. Turner (2006), 35.

36. Polletta (2006), 126, emphasis in original.

37. Ibid., 128.

38. Ibid.

39. Staff activists and the engineers did make occasional forays across borders to build stations, though these transmitters were exceptionally unwieldy and in a great enough state of disrepair that it was quite unlikely that this would be their fate. I argue that this claim was most important for its relationship to an activist imaginary, supporting community self-determination through access to media technology. See Wolfson (2012) for a critique of the activist practice of placing communication technology at the center of resistance.

40. Benkler (2006); Kreiss, Finn, and Turner (2010); see also Fish, Murillo, Nguyen, Panofsky, and Kelty (2011).

41. Pursell (1993), 630.

42. Ibid.: 635.

43. Chen (2009); Turner (2009).

44. Chen (2009).

45. Compare to Chen (2009): 12–13; Turner (2009).

46. According to Chen (2009), these problems may afflict either collectivist or bureaucratic organizations, though playing out differently according to the organizational form.

47. See Hacker (1990).

48. Interview (July 25, 2006).

49. Juris (2008), 17.

50. Turner (2006), 35–36.

51. Turner (2006), 256.

52. Of course, many peer production projects become institutionalized as they scale up and persist over time. Thanks to Biella Coleman for discussion on this point. On a related note, Barley and Kunda (2001) warn against what they call "conceptual inversion," the notion that "postbureaucratic" forms of organizing must somehow be the *opposite* of bureaucratic entities. They write, "Analysts seem to have embraced the image of a network organization largely because of its evocativeness as a metaphor for the widely held belief that firms are becoming less hierarchical and that cross-functional relationships are becoming more common" (2001, 77). They argue that the notion that "networks are not hierarchies" is overstated.

53. Kreiss, Finn, and Turner (2010): 252. This account also serves as a corrective to accounts that would explain change in techno-social formations as unidirectional, slouching toward technological or social progress, assuming that contemporary formations necessarily transcend past ones (see Shirky 2008). This last belief, even as it celebrates human potential through techno-social engagement of course reproduces technological determinism (Wyatt 2003).

54. Oudshoorn, Rommes, and Stienstra (2004), 55. Because gender, class, and race are addressed more extensive in other chapters, I leave them largely unexamined here. But a superficial read of the social identities among and dynamics between participants in the transmitter workshop reveals predictable patterns: women were more likely to be novices, technical expertise was likely to intersect with white masculinity, and both activists and hobbyists were likely to be middle-class and college educated.

Chapter 5

1. Dunbar-Hester (2011).

2. Gieryn (1983).

3. Lab studies accounts pay close attention to the craft, practice, rituals, tacit assumptions, rhetorical forms, and practice of science in order to contextualize and destabilize the scientists' "final products" and narratives of achievement, certainty, and consensus (see Latour and Woolgar 1979; Collins 1985; Traweek 1988; Jordan and Lynch 1992; Kohler 1994; Knorr Cetina 1999). Though this was a relatively low-tech work environment, at least when compared to many scientific work settings, and though quite a bit of this work was unpaid and volunteer rather than paid labor, there were significant parallels with labs and high-tech workplaces, because LPFM activism provides an environment or milieu in which participants explicitly and purposively engage with technology. (Nor, of course, is all of the work in high-tech workplaces actually high tech.).

4. Glaeser (2000), 187.

5. Abbott (1988).

6. I am using "profession" a bit loosely here: activism does not have a formalized credentialing system as professions are understood to, and its boundaries are more porous and workers less cleanly delineated than, say, medicine.

7. Becker and Carper (1956), 342.

8. Field notes (February 6, 2005). In summer 2007, Prometheus formally adopted consensus for decision making in an effort to reinforce the nonhierarchical ideals of the organization.

9. Thomas to basement (February 27, 2006).

10. Jasper to basement e-mail list (March 9, 2006).

11. Latour and Woolgar (1979), 47.

12. Ibid., 48.

13. Kim Fortun lists inscriptions produced by activists in Bhopal seeking justice after the Union Carbide disaster: "Pamphlets for journalists and students. Affidavits intended to overturn American denials of jurisdiction. Books for children. Position papers, and poems" (2001, 263).

14. Latour and Woolgar (1979), 48.

15. Prometheus Radio Project, RFP for FM Radio Interference Study (May 15, 2003), received via e-mail (June 4, 2003).

16. "Broadcast Lobby Caught Red Handed with Red Herring—'Oceans of Radio Interference' Proven to Be Puddles by Independent Study of LPFM," Prometheus Radio Project (July 13, 2003). The reference to "imaginary uranium shipments" was a pointed criticism, linking NAB's characterization of LPFM interference to the false pretenses under which the Bush administration drew the nation into an invasion of Iraq.

17. Fortun (2001), 262–263.

18. Riles (2000), 132.

19. E-mail to basement (February 13, 2006).

20. E-mail to basement (January 18, 2006).

21. Field notes (February 7, 2005).

22. Ibid.

23. Ibid.

24. Field notes (March 7, 2005).

25. Field notes (February 8, 2005).

26. Field notes (June 29, 2005).

27. Karr (2005).

28. Field notes (July 13, 2005).

29. Field notes (July 14, 2005).

30. Ibid.

31. Someone asked about Air America (the progressive talk radio network launched in 2004, which failed in 2010). Jasper replied, stating that Air America was focused on commercial stations, and mentioned Pacifica as an example of a left-wing non-commercial network. He also discussed Radio4All, an association of grassroots broadcasters and producers who exchange free content over the Internet. He called Radio4All "a poor people's A.P. [Associated Press]" (Field notes July 14, 2005).

32. Kim Fortun describes this as "recognizing that 'what counts' depends on the specific ways allegiance crosscuts any audience," a general issue in activism (2001), 262–263.

33. Field notes (July 14, 2005).

34. Ibid.

35. Epstein (1996), 3. Epstein's case of AIDS activists' interventions into AIDS research and scientific knowledge making is not analogous to the case of Prometheus in Washington because the technical issues surrounding LPFM were largely "settled," but his discussion of the boundary between expert and layperson is relevant here.

36. Field notes (March 14, 2005).

37. "We are a grassroots, collectively managed radio station dedicated to broadcasting the vibrant voices of the Knoxville community and beyond. Our very existence is a statement about our refusal to accept what is filtered down to us through mainstream TV, print and radio. We are the people, we are the media and we are dissenting from the ground up." From http://www.crok.org (February 10, 2006).

38. Incontrovertibly, cooking, cleaning, or database maintenance are also "technical" activities, but not in the sense that matters here.

39. See Schutz (1973) for the problem of first- and second-order constructs in social science.

40. E-mail, Jasper to basement (February 20, 2006).

41. E-mail, Brian to basement (February 20, 2006).

42. E-mail, Jasper to basement (February 20, 2006).

43. E-mail, — to Stubblefield (February 26, 2005).

44. E-mail, Alice to Stubblefield (May 1, 2005).

45. E-mail, — (February 25, 2005).

46. From http://lists.Prometheusradio.org/cgi-bin/mailman/listinfo/leti, (January 23, 2006).

47. E-mail, Brian to Leti (September 30, 2005).

48. This simple reference to a web page being down or otherwise unmaintained provides a reminder of the fragility of organizing and the mundane failures of organizing tools, including digital ones commonly assumed to lighten the loads of activists. See Nielsen (2009).

49. See chapter 7 for a discussion of the radio activists' attitudes toward open source software.

50. E-mail, Brian to Leti (September 9, 2005).

51. Orr (1996), 76.

52. E-mail, — to Leti (September 8, 2005).

53. E-mail, — to Leti (October 27, 2005).

54. E-mail, — to Leti (October 28, 2005). Place names have been changed.

55. E-mail, — to Stubblefield (March 2, 2006).

56. This is in some ways parallel to the conflict discussed by Steven Epstein (1995) for activists who draw initially on outsider, "lay" expertise but who become "inside" enough to participate in decision making as experts.

57. E-mail, Ellen to basement (February17, 2006).

58. E-mail, Jasper to basement (December 2, 2004).

59. E-mail, Ellen to basement (December 18, 2004).

60. A factor that this analysis does not consider in much depth is that of age, which may be important, because for some of these people, none of whom were over about thirty-five, the nonconformism of an activist identity may map well onto a performance of a youthful identity; at a staff meeting, the idea was floated that the organization should try to get a "pensioner" to volunteer, meaning someone with professional skills from an earlier career that would be useful in the office, but also someone who would be "mellower" in certain ways. And at the board meeting, a dynamic emerged between Jasper, who sometimes himself assumed a rather patriarchal role with the other organizers and volunteers, and an older board member, Alice, who was about sixty. In this case, Jasper was actively deferent toward Alice, even reminding her to discipline Prometheus as an organization; he said that as young people without professional identities, the Prometheus staff didn't know how to run the board meeting adhering to proper procedures, let alone the nonprofit.

61. Haring (2006), 8. Haring argues for a dual sense of technical identity, possessed by people and machines, mutually reinforcing and coproduced. However, I find her argument more convincing as it applies to people and thus do not take it up to explain the attribution of meaning to artifacts by people. Thanks to Cyrus Mody for a long-ago conversation about this.

62. See Coleman (2004), Jordan and Taylor (2008), Turkle (1995) on hackers; Douglas (1987), Haring (2006) on hams; and Orr (1996) on technicians.

63. See Campbell (2005). See also Klinenberg (2007), 261.

64. Jasper to basement (November 14, 2006).

65. Ibid.

66. Field notes (February 6, 2005).

67. Brian to basement (February 21, 2006).

68. The intersection of class, work, and technology requires subtle parsing. Specialized technical work is not always equated with managerial identity, or with high wages: "high-tech" office workers may identify as middle class while making relatively low wages, even as their work with computers in a multinational corporate office environment carries high status (see Freeman [2000]); other specialized technical workers may, despite higher wages, formal education, and other characteristics that might link them to the managerial class, identify more with traditional working class identities (Creighton and Hodson [1997]). Thanks to Erik Hoversten for sharing his observations and thoughts about this topic in Pasquo, Tennessee (April 2005).

69. Sociologist Eric Klinenberg writes that "by cooperating with the FCC, [Prometheus found itself] opposed to pirate leaders such as Dunifer of Free Radio Berkeley and Kantako from Human Rights Radio [unlicensed broadcasters who famously battled the FCC in court in the 1980s and 1990s]" (2007, 254).

70. Field notes (April 27, 2005).

71. Ibid.

72. Field notes (March 8, 2005).

73. Thanks to Mike Lynch for this observation.

74. Community and pirate radio in South Africa also have a special meaning given their history in resistance of the apartheid regime, which is an important element here, too: the homemade radio represented technical and overtly political values.

75. Jones (1991).

76. In some workplaces, the distinctions between work practices and tasks, as well as titles and positions, serve to differentiate; even when tasks are similar, worker differentiation occurs through identity work (Doing 2004).

77. Interview (February 16, 2006).

78. For example, Ellen to basement (January 19, 2006).

79. In an e-mail, Jasper wrote "i have some concerns that [our] organizing model is being shaped by the funding available ... my gut tells me that we will see a way that this [request] makes sense in terms of our overall strategy. so please proceed with the process. but i'd like to have a full conversation about it at some point so we are fully cognizant of what we are doing and we are not just slapping the [words donors want to hear] on what we are already doing." Jasper to basement (January 29, 2006).

Chapter 6

1. See http://uoresearch.uoregon.edu/content/story-critical-movement-sheds-light-oregon%E2%80%99s-latino-population.

2. This is possibly a reference to "Another world is possible," the motto of the World Social Forum, an annual meeting of transnational activists and NGOs to contest neoliberal corporate globalization.

3. August 20, 2006. Available at http://reclaimthemedia.org/communityradio/saludo_radialistas_a_message_from_jose_ignacio_lopez_vigil. Translated from Spanish by Lilja Otto.

4. This stands in contrast to many analyses that focus on the design and implementation stages of newer technologies.

5. Quoted in Wang (2002), 346.

6. Although the radio activists sought to contest the dominant social and political arrangements surrounding radio, they were uninterested in altering its strictly "technical" aspects. However, this is a fine line: advocating for a vastly smaller-scale artifact (100 watts of power versus 100,000 watts) constitutes a challenge to the dominant sociotechnical iteration of radio. It is impossible to strongly separate the "technical" from the "social" or "political," of course.

7. Paul Edwards argues for an elaborated notion of "discourse," holding it to be a self-elaborating "heterogeneous ensemble" that combines techniques and technologies, language, and practices (1996, 37–41), and is elaborated at multiple sites.

8. McChesney (1999), 18.

9. Ibid., 27–28.

10. Vaillant (2002), 66.

11. McCauley (2002), 518; Pickard (2011).

12. McCauley (2002), 517–518. McCauley and others note that no "chilling" economic effects of the Fairness Doctrine on stations had been proved, though this was the rationale for abandoning it.

13. Thomas Streeter sketches how commercial broadcasters constrain media content in the name of "neutrality." He argues that corporate liberalism in US broadcasting has led to "dominant corporations at the center surrounded by a periphery of smaller, more entrepreneurial firms," which do censor, "but according to the dictates of bureaucratic formulas, rarely if ever according to explicit political agendas" (1996, 38).

14. Schiller (2007), 125. See also Coyer (2005).

15. Field notes (August 18, 2006).

16. August 20, 2006. Available at http://reclaimthemedia.org/communityradio/saludo_radialistas_a_message_from_jose_ignacio_lopez_vigil. Translated by Lilja Otto. Of course the campesinos' radio station wasn't yet on the air; the activist's point about "where was your radio that day" was rhetorical.

17. Flaccus (2006).

18. Field notes (August 19, 2006).

19. Campbell (2005).

20. Field notes (August 19, 2006).

21. Ibid. Emphasis added.

22. Field notes (August 18, 2006; March 13, 2005).

23. PCUN Radio website, http://www.pcun.org/radio (July 29, 2013). Emphasis in original.

24. Tanzanian broadcasting policy allows freedom of expression in principle, but in the language set out in the policies, it makes vague recommendations that stations promote national unity and national security (Article 19: Global Campaign for Free Expression 2004, 11). In practice, broadcasters may shy away from criticizing the government too explicitly, and particularly in the semi-autonomous Zanzibar archipelago, there have been repeated instances of harassment and detainment of journalists. The community center leaders were thus extremely cautious about how to proceed with the radio station; they felt that as an independent, nonstate media outlet, they would potentially be under intense scrutiny.

25. Field notes (August 17, 2005).

26. The station's launch also coincided with Black August, a political hip-hop festival held each year in which American political hip-hop artists tour Africa, a celebration of "hip-hop and freedom fighters." Its origins are the 1970s black militant community in the California prison system. During the traditional Black August observance, observers wore black armbands, fasted during daylight, read revolutionary works, including those of George Jackson, and, interestingly, did not watch tele-

vision or listen to the radio (The Black August Hip-Hop Project website, http://mxgm.org/blackaugust).

27. Field notes (August 17, 2005).

28. [Community Center Radio Station] Rules (August 2005). This document was posted in English and Swahili in the station.

29. Historian of broadcasting Susan Smulyan insists that "not only the content of broadcasting, but its economic and technological structure ... are part of the capitalist hegemony," echoing cultural theorist Stuart Hall and Marxist theorist Antonio Gramsci (1994: 4). According to Smulyan, broadcasting did not naturally or inevitably align itself with corporate interests, but once its structure was settled on, its potential to be used in other ways was marginalized.

30. For example, Esther Dyson wrote in 1997, "The Net offers us a chance to take charge of our own lives and to redefine our role as citizens of local communities and of a global society. It also hands us the responsibility to govern ourselves, to think for ourselves, to educate our children, to do business honestly, and to work with fellow citizens to design rules we want to live by." See also Castells (1996) and Appadurai (1996) on the transformative powers of the Internet in society; Turkle (1995) for emancipatory possibilities of the Internet; as well as Marvin (1988) on the reception of new communication technologies.

31. Jasper to basement (June 9, 2006).

32. Thanks to Lucas Graves for discussion on this point.

33. Field notes (July 14, 2005). In other settings, Prometheus was rather more coy about this goal.

34. Ibid.

35. Indeed, "local" uses and "political" uses may be at odds at times; highlighting "the local" may be a way to smooth over other differences. Thanks to Greg Downey for making this point in his comments at the 2011 Society for History of Technology Meeting.

36. Informal conversation (November 5, 2007). See Downing (2003).

37. See Douglas (1987) for an excellent account of the practices of amateur operators, including DXing. Also see Smulyan (1994) and Douglas (1999).

38. Smulyan (1994), 20; Douglas (1999), 57–58.

39. Smulyan (1994), 32.

40. Slotten (2000), 26; see also Smulyan (1994), 59, and Streeter (1996).

41. Smulyan (1994), 57.

42. Ibid., 126.

43. Slotten (2000), 51.

44. Ibid., 57.

45. Streeter (1996), 106.

46. Slotten (2000), 59. According to Slotten, one commissioner, Ira Robinson, opposed the 1928 allocation plan and refused to take part in its implementation on the grounds that it favored commercial networks, but he was alone in his opposition (ibid.).

47. Streeter (1996), 59.

48. Note that Prometheus felt compelled to address the issue of the FCC's reluctance to regulate content.

49. Comments of the Prometheus Radio Project before the FCC, MM Docket No. 99–25 (July 29, 1999).

50. Comments of the United Church of Christ et al. before the FCC, MM Docket No. 99–25 (August 2, 1999).

51. Ibid.

52. Federal Communications Commission, Low Power FM Radio: An Applicant's Guide (April 28, 2000c), 5.

53. Here she is quoting the Communications Act of 1934: "Except as otherwise provided in this Act, the Commission from time to time, as public convenience, interest, or necessity requires shall—(g) Study new uses for radio, provide for experimental uses of frequencies, and generally encourage the larger and more effective use of radio in the public interest." Interview, Gloria Tristani (October 5, 2006).

54. Interview, Gloria Tristani (October 5, 2006).

55. Interview (July 19, 2006).

56. Ibid. Of course, the FCC is not monolithic in its commitment to localism. In 2006, a scandal erupted over the FCC's having buried two reports it had itself commissioned in 2003–2004 that demonstrated that local television news reporting and independent radio station ownership were damaged by media consolidation. Yet Kevin Martin, who was chairman of the FCC when the reports emerged in 2006, and Michael Powell, who was chairman when the studies were authored, both moved to promote further consolidation. See McChesney (2006).

57. Federal Communications Commission (2000a).

58. United States Senate, 109th Congress, 1st Session (February 8, 2005).

59. Interview (August 19, 2006).

60. E-mail to author (March 15, 2007).

61. Interview (August 19, 2006).

62. Field notes (February 7, 2005).

63. Houston Indymedia (2005).

64. New Orleans Indymedia (2005).

65. E-mail, — to Stubblefield (January 23, 2006).

66. E-mail, Ellen to Stubblefield (July 20, 2006). The WARN Act was approved by Congress in October 2006 but did not ultimately include language to restore or expand LPFM.

67. See Haring (2006), chapter 5.

68. Field notes (August 18, 2006).

69. Ibid. Eric Klinenberg (2007) cites a 2002 disaster in Minot, North Dakota, in which a toxic gas cloud threatened the town; none of the six local commercial radio stations, all owned by Clear Channel, could be reached to make an on-air warning to residents. One person died and more than a thousand people were injured.

70. Field notes (August 18, 2006).

71. Future of Music Coalition website (2006).

72. Interview (August 19, 2006).

73. Prometheus Radio Project website (2003).

74. Progressive churches such as the United Church of Christ have been significant advocates for media reform, including LPFM, since the 1960s (see Horwitz 1997).

75. Apostolidis (2002) argues that there is a stark contrast between the Christian right's evangelical and fundamentalist theology and its embrace of modern communications technologies, as well as older technologies like radio.

76. Quoted in Leonhardt (2000).

77. Interview (September 20, 2006).

78. Ibid.

79. Quoted in Karr (2005): 22.

80. E-mail, Ellen to author (January 19, 2005).

81. Field notes (July 14, 2005).

82. Interview (September 20, 2006).

83. Quoted in Boehlert (2000).

84. National Public Radio (1998).

85. Though the 1967 Public Broadcasting Act makes it clear that the intent is not to support locally originating *signals* piping in outside content, ambiguity over what constitutes locally originating content begins with the FCC's own definition of locally originating content, which allows remarkable leeway. This points to additional difficulty pinning down what "localism" is.

86. I argue elsewhere that "localism" is a fluid, even protean concept, less interesting to define than to trace out in terms of the uses to which actors put it in the 1990s to 2000s in debates over low-power radio. See Dunbar-Hester (2013); see also Kirkpatrick (2006).

87. Smulyan (1994), 1.

88. Rothenbuhler and McCourt (2002), 369.

89. Lloyd (2005), 89.

90. Ibid.

91. Rothenbuhler and McCourt (2002), 376.

92. Clifford (2008).

93. Coyer and Tridish (2005); National Public Radio website (http://www.npr.org/about-npr/178640915/npr-stations-and-public-media). Coyer and Tridish argue that "almost the entire US population has access to a signal that carries NPR programming, and many towns have several stations that carry NPR but no community station that carries local news" (footnote vi).

94. In the matter of reply comments for RM 9208, RM 9242, as well as MM98–93 and the Biennial Review of Broadcast Ownership, [Jasper] to the Federal Communications Commission (ca.1998).

95. Captain Avery was an English pirate who operated in the Atlantic and Indian Oceans in the 1690s.

96. United Church of Christ et al. (August 2 1999): 25.

97. In the matter of reply comments for RM 9208 ... (ca. 1998). See also Bagdikian (2000); Herman and Chomsky (1988).

98. As recently as 2005, charges of Payola dogged the commercial recording and radio industries, when New York State attorney general Elliot Spitzer forced Sony-BMG to acknowledge its participation in Payola and pay a $10 million settlement (Office of the New York State Attorney General 2005).

99. Interview (July 19, 2006).

100. All the bands he mentioned were well-known indie groups, some of whom were on major "independent" record labels such as Matador and Merge, both

founded in 1989; they are hardly obscure. See Kruse (2003) for a discussion of the indie music scene, including the contradictions between musical artists who view themselves as "independent" even while they record for major labels.

101. Interview (July 19, 2006).

102. See Keith (2002); Walker (2001) for accounts of commercial and noncommercial free-form and underground radio in the 1960s and 1970s.

103. Prometheus Radio Project, Community Radio Says: Measure Me! (n.d.).

104. Greve, Pozner, and Rao (2006): 832.

105. Smulyan (1994), 160–161.

106. Interview (July 11, 2003).

107. Interview (July 14, 2003).

108. Interview (July 10, 2003).

109. Thanks to Ron Kline for a conversation about this "anti-Arbitron" positioning.

110. Jesse Walker gives examples of community stations with mostly low-income listeners that were hurt when federal money was withdrawn (2001, 146).

111. E-mail, PRP to CPB, Comments of Prometheus Radio Project for CPB on Funding of LPFM Stations (October 13, 2004).

112. Prometheus Radio Project, Community Radio Says: Measure Me! (n.d.).

113. Ibid.

114. Interview (July 14, 2003).

115. Field notes (August 19, 2006).

116. Field notes (August 20, 2006).

117. National Association of Broadcasters (2003), 11.

118. Ibid., 12.

119. Interview (July 21, 2006).

120. "Don't hate the media, be the media" was a common refrain in Indymedia as well. See Wolfson (2014).

121. E-mail, — to Media Act list (March 14, 2007).

122. Smulyan (1994), 10; see also Streeter (1996), 302.

123. See Wolfson (2014); Carroll and Hackett (2006); Dunbar-Hester (2008).

124. See Kline (2000), conclusion.

125. Slotten (2000), 57.

126. Carroll and Hackett (2006): 95.

127. See Boczkowski and Lievrouw (2008). Beyond the issue of "content," radio, and indeed media and information technologies generally, present additional challenges to schematize, because the lines among users, consumers, mediators, and producers may be blurry. The radio activists, as propagators of radio, were producers who exported meaning and built and distributed artifacts and skills, while having an interest in which sorts of broadcasts were produced and transmitted. They were also consumers of technologies (those required for transmission and audio production, not to mention reception, because all of these actors were additionally listeners). See Thomas Misa (1995, introduction) for a discussion of ways in which the consumers of steel were also the producers, simultaneously creating demand and producing material.

128. Jordan (1999).

129. Hoynes (2002).

130. Cox (2003).

131. Media historian Joshua Greenberg contends that any technology—even toasters—may be construed as media technology if users concur that content is being mediated through the technology (2008, introduction). Of course, technologies *not* understood to be media technologies may also present the analyst with something resembling "content," for example, aesthetic or semiotic value held to be "transmitted" or ascribed to them by users. (See also Sterne [2006] for a discussion of the "container" metaphor.)

132. Carroll and Hackett (2006), 95.

133. The radio activists' emphasis on the suitability of radio for localism rather echoes Marshall McLuhan's statement that radio "is a decentralizing, pluralistic force," with which they would undoubtedly agree (1964, 267). But whereas the radio activists saw this as leading to more democratic conditions, McLuhan attended to radio's impact in its use by both Irish rebels and Hitler. See McLuhan (1964), 259–268.

Chapter 7

1. Fouché (2006), 654; see also Shields and Ogles (1995).

2. Field notes (February 24, 2006).

3. Ibid.

4. Ibid.

5. Emerging technologies are often viewed through the lens of patterns of use and interpretation of older technologies, at least initially (Boczkowski 2004, 3).

6. Writing of newspapers' halting and uneven adoption of digital media, Pablo Boczkowski says, "In contrast with the discourse about revolutionary effects that have been prevalent in the dominant modes of understanding online technologies and the web, ... innovations [unfold] in a more gradual and ongoing fashion ... shaped by various combinations of initial conditions and local contingencies" (2004, 4).

7. Field notes (March 1, 2005).

8. Field notes (March 3, 2005).

9. See for example Smulyan (1994); Streeter (1996); Douglas (1999); McChesney (1999); Slotten (2000).

10. Field notes (February 6, 2006).

11. "Radio" has traditionally referred to telephonic broadcasting that uses the electromagnetic spectrum, though the expansion of this term to encompass Internet radio (also known as webcasting) clearly privileges the telephonic, aural qualities and downplays the means of transmission and use of the spectrum. Terrestrial radio means that it is broadcast from antennas on the ground as opposed to from satellites. Traditional terrestrial broadcasting has used analog transmission and reception technologies. The use of digital signals is a more efficient use of spectrum because more channels can be fit into the same amount of bandwidth than with analog transmission, but analog transmission and reception are well established, and digital signals have been criticized for having less depth and richness in terms of sound than analog. Digital signals are used in satellite broadcasting, and the conversion of the FM spectrum to digital (which will provide additional channels and necessitate different transmission and reception technologies) has been a controversial topic as it was considered by the FCC and other groups; a digital rollout is anticipated in the future. Here, FM and LPFM are understood as analog, terrestrial radio, without much consideration of the radio activists' attitude toward digital radio, because it was not a main focus of their activities. To the extent that they did consider it, the activists discouraged LPFMs from being early adopters of digital, instead encouraging them to wait until groups with more money and a bigger stake finished wrangling over specifications (field notes March 16, 2005), and they regarded with suspicion the suggestion that they wait for more spectrum to be parceled out, with more frequencies becoming available when the transition to digital occurs. One activist said, "Digital broadcasters are under no obligation to serve the public interest [unless we make them]" (field notes March 3, 2005).

12. Field notes (February 6, 2005).

13. "Access" might be viewed as doing discursive work similar to "localism": no one could oppose it. But arguably, "access" is an impoverished concept that detracts

from notions of greater potential. Activists both employed this discourse and critiqued it at times. Thanks to Steve Jackson for comment on this; also see Eubanks (2011).

14. Field notes (February 6, 2005).

15. Interview (July 5, 2005).

16. McChesney (2005), 17.

17. Walker (2001), 214.

18. Field notes (March 16, 2005).

19. Interview (July 3, 2003). This activist also said that "webcasting ... it's getting better, but the comic book guy from *The Simpsons* [television show], that's your target audience." "Comic book guy" is a social misfit and a geek, known for holing up to watch marathons of the television program *Dr. Who*; I take this activist's comment to mean that he understood comic book guy to be an elite user of technology concerned with arcane topics, not a likely participant in community media or community technology projects. Thanks to Cyrus Mody for discussion on this point.

20. Interview (June 19, 2003).

21. Interview (July 14, 2003).

22. Interview (July 5, 2006).

23. Walker (2001), 278.

24. Terrestrial broadcasters are not required to pay to play songs on the radio, but webcasters must pay an annual fee (which was set to increase in 2007, generating controversy), as dictated by the Copyright Royalty Board, a board of judges appointed by the Library of Congress.

25. Field notes (March 16, 2005).

26. Code of Federal Regulations, 47 CFR §15.

27. See Sandvig (2005) on the regulatory expansion of the "open" use of spectrum.

28. I use *wi-fi* and *wireless* more or less interchangeably. *Wi-Fi* was a brand name but has broadened to refer to a family of related specifications (the IEEE 802.11 group), which specify methods and techniques of wireless local area network operation. *Wi-fi* is a play on the term *hi-fi* but does not actually stand for "wireless fidelity."

29. Wi-fi was initially intended to network homes or businesses, but increasingly it has been seen as a means by which to network municipal areas (Tapia, Maitland, and Stone 2006: 362).

30. Meinrath (2005), 230–236.

31. Ibid., 228.

32. Note again the reference to Appropriate Technology, as detailed in chapters 2 and 4.

33. Field notes (February 23, 2006).

34. Interview (June 29, 2006).

35. Interview (September 26, 2006). See also Streeter (1996).

36. Interview (June 29, 2006).

37. Field notes (February 23, 2006).

38. These community wi-fi networks strongly resemble the independent telephone companies formed in the era of early telephony, built by residents of small towns and rural areas. See Sandvig, Young, and Meinrath (2004): 4; see also Fischer (1992); Kline (2000).

39. From FreePress.net. Available at http://www.freepress.net/communityinternet/=states (June 14, 2007). See Tapia, Maitland, and Stone (2006) for an overview of municipal wi-fi.

40. Field notes (January 13, 2005). See also Meinrath (2005), 225.

41. Interview (June 29, 2006).

42. Ibid.

43. Field notes (March 2, 2005).

44. Ibid.

45. Thanks to Javier Lezaun for conversations about this topic.

46. See also Hayes (2000), who draws on Roland Barthes for her discussion of the qualities of "voice." Ihde (1976) also discusses the characteristics of sound, listening, and voice from a phenomenological perspective. I do not argue that the "voice" is an essential characteristic of an aural medium, rather that the notion of "voice" is associated with discourse about democratic participation. However, radio is fundamentally an aural medium, whereas computers use a visual interface.

47. Field notes (January 13, 2005).

48. See Kvasny (2005); Eubanks (2011) for excellent critiques of "the digital divide" and "digital inclusion."

49. Field notes (March 10, 2006).

50. What Is Community Wireless? (n.d.). Received spring 2006. Emphasis added.

51. E-mail to basement (April 25, 2006).

52. Meinrath (2005), 228.

53. Neighborhood Access to Technology. Wireless Community Networks User's Guide (nd). Distributed March 2005.

54. Media Tank. Our Neighborhoods Need Access Today! (nd). Emphasis in original.

55. Field notes (March 3, 2005).

56. Field notes (March 1, 2005).

57. Fouché (2006): 642. See also Hampton (2004).

58. Field notes (March 1, 2005).

59. Interview (September 26, 2006).

60. Field notes (March 2005).

61. Fouché (2006): 647.

62. Field notes (March 1, 2005).

63. In using this term, I invoke the insight provided by critical whiteness studies, viewing race as a historical, political category that constructs and privileges whiteness in relation to and at the expense of nonwhites. See Hill (1997).

64. Some tension about "giving" emerges in the following statement by an organizer: "Prometheus wants to encourage US LPFM people to build/bring transmitters and radio technology whenever they travel. We should establish partnerships and sister station relationships elsewhere and give back to the rest of the world so the movement grows and we [Global North countries, particularly the United States] are not resented. The barrier elsewhere isn't the energy, it's access to stuff [technological artifacts]"(field notes March 16, 2005).

65. The activists' use of racial categories should also be read as being in dialogue with the contributions of critical whiteness and antiracism. For this group of activists, the need for media activism to ally itself with antiracism is not an afterthought; as noted elsewhere, many drawn to media activism began activist work on other social justice causes, but then identified media justice as the linchpin of their advocacy (Carroll and Hackett 2006; Dunbar-Hester 2008). Thus, much media activism can be seen as a means to an end in a larger struggle against inequality. That said, media activism is not immune to the problems faced in other social movements, which struggle with inclusion, representation, and differences in privilege and positionality among members of the movement. See for example hooks (1989) on the adverse effects of racism on the women's movement.

66. Interview (February 16, 2006).

67. Interview (September 26, 2006).

68. The activists held two of approximately ten radio station barnraisings between 2003 and 2008 with Latino migrant workers' unions, in 2003 and 2006 (Immokalee, Florida, and Woodburn, Oregon).

69. Informal conversation (June 28, 2007).

70. Field notes (February 6, 2005).

71. Field notes (March 1, 2005).

72. Field notes (March 16, 2005).

73. See Coleman (2012); Kelty (2008).

74. Interview (July 14, 2003).

75. Ibid. In the intervening years since this interview, it is increasingly evident that hacker politics range widely. See Coleman (2004, 2012).

76. See Sandvig, Young, and Meinrath (2004); Meinrath (2005); Söderberg (2011).

77. See Abbate (1999); Sandvig, Young, and Meinrath (2004).

78. What Is Community Wireless? (nd). Received spring 2006.

79. Sandvig, Young, and Meinrath (2004): 12.

80. Engineering culture itself provides some obstacles to the promotion of sharing, according to Christian Sandvig (2005: 18).

81. Carolyn Marvin notes that marginalization of the public from engineering expertise is as old a practice as the profession itself (1988: chapter 1).

82. Interview (September 26, 2006).

83. Ibid.

84. Bozckowski (2004), 3. See also Bolter and Grusin (2000); Wyatt (2003).

85. My account of mediation complements and extends user-focused analyses (see Woolgar 1991; Akrich 1995; Kline and Pinch 1996; Oudshoorn and Pinch 2003; Fouché 2006). Of course, the distinction between mediators and users is not always clear-cut. Boczkowski (2004) details the negotiation of electronic options in print newsrooms; his "users" are not the end users. This underscores what a complex category "the user" may be, as well as how this category is perhaps particularly difficult to theorize in the case of communications technologies.

86. Cravens, Dailey, and Wallace (2007). Sandvig (2005) also makes this association.

87. See Eglash and Bleecker (2001); Eglash et al. (2004).

88. Field notes (March 13, 2005).

89. Eglash and Bleecker (2001); Fouché (2005); Kvasny (2005).

90. Field notes (February 6, 2005). See Haraway (1991).

91. Informal conversation (November 5, 2007).

92. One day in the office, an activist complained, "My worldview is that computers should adapt to the way you think; you shouldn't have to adapt to the way it thinks," again demonstrating some ambivalence toward computers (field notes November 17, 2004).

93. Marvin (1988), 8.

Conclusion

1. Waits (2012).

2. Bagdikian (2000); Herman and Chomsky (1988); Klinenberg (2007); McChesney (1999); McChesney et al. (2005); see also Starr (2004).

3. Streeter (2011), 3.

4. Here I draw on the work of feminist scholars of science and technology such as Donna Haraway (1991) and Lucy Suchman (2003). See also Dunbar-Hester (2014b) for more on problems of universality in technical DIY.

5. Eglash et al. (2004), xv; Eubanks (2011).

6. See Marvin (1988). Thanks to Kristen Haring for this phrasing.

7. Pinch and Bijker (1987); Rosen (1993); Kline and Pinch (1996); Oudshoorn and Pinch (2003).

8. See Edgerton (2006) and Jackson (2013) for more on the vast work of maintenance and repair of old technologies.

9. Horwitz (1997); McChesney (1999); Pickard (2011).

10. A related point is that what we communicate about technology reflects the priorities and concerns of the society, revealing as much about the communicators as about the artifacts in question (Sturken and Thomas 2004).

11. See Shannon (1948); Shannon and Weaver (1949); Schiller (1988); Winner (1988); Bowker (1994); Webster (2002); Kline (2004, 2006).

12. Greenberg writes that "there is no inherent reason we should [analytically] treat someone watching a television differently from someone watching a hammer or a fork, but we do" (2008, 9); see also McLuhan (1964, 28) on content and message; Lievrouw and Livingstone quoted in Boczkowski (2004, 11). Alternate, overlapping terms for "media technology" include "communication technology" and "information technology." I tend to use "media technology" and "communication technology" fairly interchangeably, because actors also merge these categories, taking these

technologies to be components of "the media," a larger set of institutions, networks, and artifacts than is indicated by technological artifacts alone (Starr 2004).

13. Kline (2006); Williams (1976). Kline's analysis is limited to the use of the term by professional groups, but this analysis, of a later historical period, includes policy experts, activists, and members of community groups.

14. Kline (2006); Marx 2010.

15. Douglas (1987), 303; see also Carey (1989), 18.

16. Quoted in Wells (2006). Emphasis in original.

17. Winner (1988).

18. National Public Radio (1998).

19. Ibid.

20. This valence of information may relate in part to Claude Shannon's (1948) insistence on bracketing out meaning or semantics in his technical, mathematical definition of information. See Hayles (1999) for an analysis of how information came to be disembodied and Turner (2006, 38) for origins of this understanding of information in the countercultural deployment of cybernetics.

21. Williams also discusses the relation of "community" to the German concept of *Gemeinschaft* and the French notion of the commune, the smallest administrative division; "the sense of immediacy or locality was strongly developed in the context of larger and more complex industrial societies" Williams (1976, 75–76); see also Tönnies (2001).

22. Turner (2006, chapter 5).

23. This is in no way to claim that this "freedom" has been achieved in any of the cases it has been claimed: my aim is only to point out the reach and success of the discourse.

24. Informal conversation (November 5, 2007).

25. Field notes (July 2006).

26. Straus (2007).

27. To take this position, one need not accept that technology is a "neutral tool"—equally capable of being put to "good" or "evil" purposes—and this is certainly not what I am advocating (Winner 1988).

Bibliography

Abbate, Janet. *Inventing the Internet*. Cambridge, MA: MIT Press, 1999.

Abbott, Andrew. *The System of Professions: An Essay on the Division of Expert Labor*. Chicago: University of Chicago Press, 1988.

Akrich, Madeleine. User Representations: Practices, Methods, and Sociology. In *Managing Technology in Society*, ed. Arie Rip, Thomas Misa, and Johan Schot. London: Pinter, 1995.

Anderson, Benedict. *Imagined Communities*. New York: Verso, 1991.

Anderson, John. A Can of Worms: Public Intransigence on the Public Airwaves. Master's thesis, University of Wisconsin (2004).

Antonijevic, Smiljana, Stefan Dormans, and Sally Wyatt. Working in Virtual Knowledge: Affective Labor in Scholarly Collaboration. Working paper, Virtual Knowledge Studio for the Humanities and Social Sciences, Amsterdam, 2010.

Appadurai, Arjun. *Modernity at Large*. Minneapolis: University of Minnesota Press, 1996.

Apostolidis, Paul. Scanning the "Stations of the Cross": Christian Right Radio in Post-Fordist Society. In *The Radio Studies Reader*, ed. Michele Hilmes and Jason Loviglio. New York: Routledge, 2002.

Arendt, Hannah. *The Promise of Politics*. New York: Schocken, 2005.

Article 19: Global Campaign for Free Expression. Note on the United Republic of Tanzania Information and Broadcasting Policy (February 2004). Available at http://www.article19.org/pdfs/analysis/tanzania-information-and-broadcasting-policy-f.pdf.

Bagdikian, Ben. *The Media Monopoly*. 2nd ed. Boston: Beacon Press, 2000.

Barley, Stephen, and Gideon Kunda. Bringing Work Back In. *Organization Science* 12 (2001): 76–95.

Barnes, Barry, and David Bloor. Relativism, Rationalism, and the Sociology of Knowledge. In *Rationality and Relativism*, ed. Martin Hollis and Steven Lukes. Cambridge, MA: MIT Press, 1981.

Becker, Howard. *Outsiders: Studies in the Sociology of Deviance*. New York: Free Press, 1963.

Becker, Howard. *Art Worlds*. Berkeley: University of California Press, 1982.

Becker, Howard, and James Carper. The Elements of Identification with an Occupation. *American Sociological Review* 21 (1956): 341–348.

Benkler, Yochai. *The Wealth of Networks: How Production Networks Transform Markets and Freedom*. New Haven, CT: Yale University Press, 2006.

Bijker, Wiebe. The Social Construction of Fluorescent Lighting. In *Shaping Technology/Building Society*, ed. Wiebe Bijker and John Law. Cambridge, MA: MIT Press, 1994.

Bijker, Wiebe, Thomas P. Hughes, and Trevor J. Pinch. *The Social Construction of Technological Systems*. Cambridge, MA: MIT Press, 1987.

Bimber, Bruce. Three Faces of Technological Determinism. In *Does Technology Drive History? The Dilemma of Technological Determinism*, ed. Merritt Roe Smith and Leo Marx. Cambridge, MA: MIT Press, 1994.

Black August Hip-Hop Project website (February 28, 2007). Available at http://www.blackaugust.com/2006/about_ba.html.

Blake, Art. Audible Citizenship and Audiomobility: Race, Technology, and CB Radio. *American Quarterly* 63 (2011): 531–553.

Boczkowski, Pablo. *Digitizing the News: Innovation in Online Newspapers*. Cambridge, MA: MIT Press, 2004.

Boczkowski, Pablo, and Leah Lievrouw. Bridging STS and Communication Studies: Scholarship on Media and Information Technologies. In *New Handbook of Science and Technology Studies*, ed. E. J. Hackett, Olga Amsterdamska, Michael E. Lynch, and Judy Wajcman. Cambridge, MA: MIT Press, 2008.

Boehlert, Eric. Mixed Signals. Salon.com (April 11, 2000). Available at http://www.salon.com/news/feature/2000/04/11/radio.

Bolter, Jay David, and Richard Grusin. *Remediation: Understanding New Media*. Cambridge, MA: MIT Press, 2000.

Bourdieu, Pierre. *Distinction: A Social Critique of the Judgment of Taste*. London: Routledge, 1986.

Bowker, Geoffrey. Information Mythology: The World of/as Information. In *Information Acumen: The Understanding and Use of Knowledge in Modern Business*, ed. Lisa Bud-Frierman. London: Routledge, 1994.

Boyer, Dominic. The Social Context of Critical Intellectual Agency: The Shifting Fortunes of the German Educated Bourgeoisie and the Criticism of Modern Society. Presented to the Society for the Humanities, Cornell University (May 2003).

Boyer, Dominic, and Ulf Hannerz. Introduction: Worlds of Journalism. *Ethnography* 7 (2006): 5–17.

Boyle, Deirdre. *Subject to Change: Guerrilla Television Revisited.* New York: Oxford University Press, 1997.

Brand, Keith. The Rebirth of Low-Power FM Broadcasting. *U.S. Journal of Radio Studies* 11 (2004): 153–168.

Brecht, Bertolt. The Radio as an Apparatus of Communication. In *Radiotext(e),* ed. Neil Strauss. New York: Semiotext(e), 1993.

Butler, Judith. *Gender Trouble.* New York: Routledge, 1990.

Butler, Judith. *Bodies That Matter.* New York: Routledge, 1993.

Butler, Judith. Against Proper Objects. *differences: A Journal of Feminist Cultural Studies* 6 (1994): 1–26.

Callon, Michael, and Vololona Rabeharisoa. The Involvement of Patients' Associations in Research. *International Social Science Journal* 171 (2002): 57–65.

Campbell, Duncan. Farmworkers Win Historic Deal after Boycotting Taco Bell. *The Guardian* (March 12, 2005). Available at http://www.Guardian.Co.Uk/Usa/Story/0,12271,1436022,00.html.

Carey, James. *Communication as Culture.* Winchester, MA: Unwin Hyman, 1989.

Carroll, William, and Robert Hackett. Democratic Media Activism through the Lens of Social Movement Theory. *Media Culture & Society* 28 (2006): 83–104.

Casper, Monica. Feminist Politics and Fetal Surgery: Adventures of a Research Cowgirl on the Reproductive Frontier. *Feminist Studies* 23 (1997): 232–262.

Castells, Manuel. *The Information Age: Economy, Society, and Culture; The Rise of the Network Society.* Vol. 1. Oxford: Blackwell Publishers, 1996.

CBS News. Powell: Jackson Flash "A New Low": FCC Chair Slams Halftime Show, but Lawmakers Say FCC Shares Blame (February 11, 2004). Available at http://www.cbsnews.com/stories/2004/02/11/entertainment/main599759.shtml.

Champion, Sarah. Interview with Lawrie Hallett, the Man from OFCOM. Community FM. Available at http://www.communityfm.net/story94.htm.

Chen, Katherine. *Enabling Creative Chaos: The Organization Behind the Burning Man Event.* Chicago: University of Chicago Press, 2009.

Clifford, James. Introduction: Partial Truths. In *Writing Culture,* ed. James Clifford and George Marcus, 1–26. Berkeley: University of California Press, 1986.

Clifford, Stephanie. Radio's Revenue Falls Even as Audience Grows. *New York Times* (November 25, 2008). Available at http://www.nytimes.com/2008/11/26/business/media/26adco.html.

Cockburn, Cynthia. The Material of Male Power. In *The Social Shaping of Technology.* 2nd ed., ed. Donald Mackenzie and Judy Wajcman. Philadelphia: Open University Press, 1999.

Coleman, E. Gabriella. The Political Agnosticism of Free and Open Source Software and the Inadvertent Politics of Contrast. *Anthropological Quarterly* 77 (2004): 507–519.

Coleman, E. Gabriella. *Coding Freedom.* Princeton, NJ: Princeton University Press, 2012.

Collins, H. M. Researching Spoonbending: Concepts and Practice of Participatory Fieldwork. In *Social Researching: Politics, Problems, Practice,* ed. Colin Bell and Helen Roberts. New York: Routledge, 1984.

Collins, H. M. *Changing Order: Replication and Induction in Scientific Practice.* London: Sage, 1985.

Collins, H. M., and Robert Evans. The Third Wave of Science Studies: Studies of Expertise and Experience. *Social Studies of Science* 32 (2002): 235–296.

Connell, R. W. *Masculinities.* Sydney: Allen & Unwin, 2005.

Coopman, Ted. Sailing the Spectrum from Pirates to Micro Broadcasters: A Case Study of Micro Broadcasting in the San Francisco Bay Area. Master's thesis, San Jose State University (1995).

Coopman, T. M. FCC Enforcement Difficulties with Unlicensed Micro Radio. *Journal of Broadcasting & Electronic Media* 43 (1999): 582–602.

Cowan, Ruth Schwartz. The Consumption Junction: A Proposal for Research Strategies in the Sociology of Technology. In *The Social Construction of Technological Systems,* ed. Wiebe Bijker, Thomas P. Hughes, and Trevor J. Pinch. Cambridge, MA: MIT Press, 1987.

Cox, Craig. *Storefront Revolution: Food Co-ops and the Counterculture.* New Brunswick, NJ: Rutgers University Press, 1994.

Cox, Ted. The FCC and Media Ownership: Where Do We Go from Here? *Chicago Daily Herald* (June 12, 2003).

Coyer, Kate. It's Not Just Radio. PhD diss., Goldsmiths College, University of London (2005).

Coyer, Kate, and Pete Tridish. A Radio Station in Your Hands Is Worth 500 Channels of Mush! The Role of Community Radio in the Struggle against Corporate Domination of Media. In *News Incorporated: Corporate Media Ownership and Its Threat to Democracy*, ed. Elliot Cohen. Amherst, NY: Prometheus Books, 2005.

Cravens, Vikki, Dharma Dailey, and Antuan Wallace. Can We Build a Wireless Communications Infrastructure That Values Everyone's Right to Communicate? © Prometheus Radio Project (2007). Available at http://www.prometheusradio.org/node/39.

Creighton, Sean, and Randy Hodson. Whose Side Are They On? Technical Workers and Management Ideology. In *Between Craft and Science: Technical Work in U.S. Settings*, ed. Stephen Barley and Julian Orr. Ithaca, NY: Cornell University Press, 1997.

Cruickshank, Douglas. Rebels on the Air review. Salon.com (February 20, 2002). Available at http://dir.salon.com/story/books/review/2002/02/11/radio/index.html.

Cyril, Malkia. Justice by Any Name. Free Press and Media Reform Conference Opening Plenary Speech, St. Louis, MO (May 13, 2005). Available at http://www.youthmediacouncil.org/publications.html.

Damarin, Amanda. Fit, Flexibility, and Connection: Organizing Employment in Emerging Web Labor Markets, New York City 1993–2003. PhD diss., Columbia University (2004).

David, Shay. Open Systems in Practice and Theory: The Social Construction of Participatory Information Networks. PhD diss., Cornell University (2008).

Dawson, Max. Home Video and the "TV Problem": Cultural Critics and Technological Change. *Technology and Culture* 48 (2007): 524–549.

de Laet, Marianne, and Annemarie Mol. The Zimbabwe Bush Pump: Mechanics of a Fluid Technology. *Social Studies of Science* 30 (2000): 225–263.

Deleuze, Gilles, and Felix Guattari. *A Thousand Plateaus: Capitalism and Schizophrenia*. Minneapolis: University of Minnesota Press, 1987.

DeNardis, Laura. *Protocol Politics: The Globalization of Internet Governance*. Cambridge, MA: MIT Press, 2009.

Dick, Steven, and Walter McDowell. Pirates, Pranksters, and Prophets: Understanding America's Unlicensed "Free" Radio Movement. *Journal of Radio Studies* 7 (2000): 329–341.

Digby, Tom. *Men Doing Feminism (Thinking Gender)*. New York: Routledge, 1998.

Doing, Park. "Lab Hands" and the "Scarlet O": Epistemic Politics and (Scientific) Labor. *Social Studies of Science* 34 (2004): 299–323.

Dornfeld, Barry. *Producing Public Television, Producing Public Life*. Princeton, NJ: Princeton University Press, 1998.

Dougherty, John. What's Peculiar about a Barnraising and Why I Like It. *Prometheus De-Livered* (Summer 2005).

Douglas, Susan. *Inventing American Broadcasting, 1899–1922*. Baltimore: Johns Hopkins University Press, 1987.

Douglas, Susan. Listening. In *Radio and the American Imagination*. New York: Times Books, 1999.

Downing, John. *Radical Media: The Political Experience of Alternative Communication*. Cambridge, MA: South End Press, 1984.

Downing, John. *Radical Media: Rebellious Communication and Social Movements*. Cambridge, MA: South End Press, 2000.

Downing, John. The Independent Media Center Movement and the Anarchist Socialist Tradition. In *Contesting Media Power: Alternative Media in a Networked World*, ed. Nick Couldry and James Curran. Lanham, MD: Rowman & Littlefield, 2003.

Dunbar-Hester, Christina. Geeks, Meta-Geeks, and Gender Trouble: Activism, Identity, and FM Radio. *Social Studies of Science* 38 (2008): 201–232.

Dunbar-Hester, Christina. "Free the Spectrum!" Activist Encounters with Old and New Media Technology. *New Media & Society* 11 (2009): 221–240.

Dunbar-Hester, Christina. Beyond "Dudecore"? Challenging Gendered and "Raced" Technologies through Media Activism. *Journal of Broadcasting & Electronic Media* 54 (2010): 121–135.

Dunbar-Hester, Christina. Drawing and Effacing Boundaries in Contemporary Media Democracy Work. In *Media and Social Justice*, ed. Sue Curry Jansen, Jefferson Pooley, and Lora Taub-Pervizpour, 195–209. London: Palgrave Macmillan, 2011.

Dunbar-Hester, Christina. Soldering toward Media Democracy: Technical Practice as Symbolic Value in Radio Activism. *Journal of Communication Inquiry* 36 (2) (2012): 149–169.

Dunbar-Hester, Christina. What's Local? Localism as a Discursive Boundary Object in Low-Power Radio Policymaking. *Communication, Culture & Critique* 6 (4) (2013): 502–524.

Dunbar-Hester, Christina. Producing "Participation"? The Pleasures and Perils of Technical Engagement in Radio Activism. *Public Culture* (Winter 2014).

Dunbar-Hester, Christina. Radical Inclusion? Locating Accountability in Technical DIY. In *DIY Citizenship: Critical Making and Social Media*, ed. Matt Ratto and Megan Boler, 75–88. Cambridge, MA: MIT Press, 2014.

Duncan, Kate. Microbroadcasting. *Z Magazine* (July–August 1998): 40–41.

Dyson, Esther. *Release 2.0*. New York: Broadway, 1997.

Edgerton, David. *The Shock of the Old: Technology and Global History since 1900*. Oxford: Oxford University Press, 2006.

Edwards, Paul. The Army and the Microworld: Computers and the Politics of Gender Identity. *Signs* 16 (1990): 102–127.

Edwards, Paul. *The Closed World*. Cambridge, MA: MIT Press, 1996.

Eglash, Ron. Race, Sex, and Nerds: From Black Geeks to Asian American Hipsters. *Social Text* 71 (2002): 49–64.

Eglash, Ron, and Julian Bleecker. The Race for Cyberspace: Information Technology in the Black Diaspora. *Science as Culture* 10 (2001): 353–374.

Eglash, Ron, Jennifer Crossiant, Giovanna Di Chiro, and Rayvon Fouché. *Appropriating Technology: Vernacular Science and Social Power*. Minneapolis: University of Minnesota Press, 2004.

Epstein, Steven. The Construction of Lay Expertise: AIDS Activism and the Forging of Credibility in the Reform of Clinical Trials. *Science, Technology & Human Values* 20 (1995): 408–437.

Epstein, Steven. *Impure Science: AIDS, Activism, and the Politics of Knowledge*. Berkeley: University of California Press, 1996.

Epstein, Steven. Activism, Drug Regulation, and the Politics of Therapeutic Evaluation in the AIDS Era: A Case Study of DDC and the "Surrogate Markers" Debate. *Social Studies of Science* 27 (1997): 691–726.

Eubanks, Virginia. *Digital Dead End: Fighting for Social Justice in the Information Age*. Cambridge, MA: MIT Press, 2011.

Federal Communications Commission. FCC Chairman Responds to House Vote to Cut the Number of Community Radio Stations by 80% (April 13, 2000a).

Federal Communications Commission. In the Matter of Creation of Low Power Radio Service. Report and Order. MM Docket 99–25, FCC 00–19 (January 27, 2000b) .

Federal Communications Commission. Low Power FM Radio: An Applicant's Guide (April 28, 2000c). Available at http://www.fcc.gov/mb/audio/lpfm/lpfmguide.pdf.

Federal Communications Commission. Statement of Dale Hatfield, Chief, Office of Engineering and Technology, and Roy Stewart, Chief, Mass Media Bureau, Concerning Low Power FM. *Engineering Issues* (March 2000d): 24.

Fischer, Claude. *America Calling: A Social History of the Telephone to 1940*. Berkeley: University of California Press, 1992.

Fish, Adam, Luis F.R. Murillo, Lilly Nguyen, Aaron Panofsky, and Christopher M. Kelty. Birds of the Internet: Towards a Field Guide to the Organization and Governance of Participation. *Journal of Cultural Economics* 4 (2011): 157–187.

Flaccus, Gillian. Spanish-Language Media Credited on Pro-Immigrant Rallies. *Boston Globe* (March 29, 2006). Available at http://www.Boston.com/news/nation/articles/2006/03/29/spanish_language_media_credited_on_pro_immigrant_rallies.

Fortun, Kim. *Advocacy after Bhopal: Environmentalism, Disaster, New Global Orders*. Chicago: University of Chicago Press, 2001.

Fouché, Rayvon. *Black Inventors in the Age of Segregation*. Baltimore: Johns Hopkins University Press, 2005.

Fouché, Rayvon. Say It Loud, I'm Black and I'm Proud: African Americans, American Artifactual Culture, and Black Vernacular Technological Creativity. *American Quarterly* 58 (2006): 639–661.

Freeman, Carla. *High Tech and High Heels in the Global Economy: Women, Work, and Pink-Collar Identities in the Caribbean*. Durham, NC: Duke University Press, 2000.

Free Press website. What We Do. Available at http://www.freepress.net/content/about.

Future of Music Coalition website. Radio Consolidation Hurts Public, New Data Shows (December 8, 2006). Available at http://www.futureofmusic.org/news/maradiostudy06.cfm.

Gangadharan, Seeta Peña. Building the Case for Change: Knowledge Practices of the Media Reform and Media Justice Movements. In *Communicating for Social Impact: Engaging Communication Theory, Research, and Pedagogy*, ed. L. Harter and M. Dutta, 161–174. Cresskill, NJ: Hampton Press, 2009.

Gattuso, James L. Raising Revenues with the Auction Option for the Telecommunications Spectrum. *Issue Bulletin* (147) (May 11, 1989). Available at http://www.heritage.org/Research/GovernmentReform/IB147.cfm.

Geertz, Clifford. *The Interpretation of Cultures*. New York: Basic Books, 1973.

Gelber, Steven. Do-It-Yourself: Constructing, Repairing, and Maintaining Domestic Masculinity. *American Quarterly* 49 (1997): 66–112.

Gergen, Kenneth. *The Saturated Self: Dilemmas of Identity in Contemporary Life*. New York: Basic Books, 1991.

Getzinger, Günter. Trans-disciplinary Research and Sustainable Technology Design. Presented at the 6th Annual Conference of the Institute for Advanced Studies on Science, Technology and Society, Graz, Austria (May 25, 2007).

Gieryn, Thomas. Boundary-Work and the Demarcation of Science from Non-Science: Strains and Interests in Professional Ideologies of Scientists. *American Sociological Review* 48 (1983): 781–795.

Gillespie, Tarleton. *Wired Shut: Copyright and the Shape of Digital Culture.* Cambridge, MA: MIT Press, 2007.

Ginsburg, Faye. Procreation Stories: Reproduction, Nurturance, and Procreation in Life Narratives of Abortion Activists. *American Ethnologist* 14 (1987): 623–636.

Gitlin, Todd. *The Whole World Is Watching: Mass Media in the Making and Unmaking of the New Left.* Berkeley: University of California Press, 1981.

Glaeser, Andreas. *Divided in Unity: Identity, Germany, and the Berlin Police.* Chicago: University of Chicago Press, 2000.

Goffman, Erving. *The Presentation of Self in Everyday Life.* New York: Anchor Books, 1959.

Goldstein, Carolyn. Part of the Package: Home Economists in the Consumer Products Industries, 1920–1940. In *Rethinking Home Economics: Women and the History of a Profession,* ed. Sarah Stage and Virginia B. Vincenti. Ithaca, NY: Cornell University Press, 1997.

Greenberg, Joshua. *From Betamax to Blockbuster.* Cambridge, MA: MIT Press, 2008.

Greve, Henrich, Jo-Ellen Pozner, and Hayagreeva Rao. Vox Populi: Resource Partitioning, Organizational Proliferation, and the Cultural Impact of the Insurgent Microradio Movement. *American Journal of Sociology* 112 (2006): 802–837.

Grint, Keith. *Work and Society.* Cambridge, UK: Polity Press, 2000.

Gusterson, Hugh. *Nuclear Rites: A Weapons Laboratory at the End of the Cold War.* Berkeley: University of California Press, 1996.

Habermas, Jürgen. *The Structural Transformation of the Public Sphere: An Inquiry into a Category of Bourgeois Society.* Cambridge, UK: Polity Press, 1989.

Hacker, Sally. *"Doing It the Hard Way": Investigations of Gender and Technology.* Boston: Unwin Hyman, 1990.

Hall, Stuart. Encoding/Decoding. In *Culture, Media, Language: Working Papers in Cultural Studies, 1972–79,* ed. Centre for Contemporary Cultural Studies. London: Hutchinson, 1980.

Halleck, Dee Dee. *Hand-Held Visions: The Impossible Possibilities of Community Media.* New York: Fordham University Press, 2002.

Hampton, Samuel. Cultural Paths to Computing: African American Women in a Community Technology Center. In *Appropriating Technology: Vernacular Science and Social Power,* ed. Ron Eglash, Jennifer Crossiant, Giovanna Di Chiro, and Rayvon Fouché. Minneapolis: University of Minnesota Press, 2004.

Hannerz, Ulf. Other Transnationals: Perspectives Gained from Studying Sideways. *Paideuma* 44 (1998): 109–123.

Håpnes, Tove, and Knut Sørenson. Competition and Collaboration in Male Shaping of Computing: A Study of Norwegian Hacker Culture. In *The Gender-Technology Relation: Contemporary Theory and Research*, ed. Keith Grint and Rosalind Gill. New York: Taylor & Francis, 1995.

Haraway, Donna. *Simians, Cyborgs and Women: The Reinvention of Nature*. New York: Routledge, 1991.

Haring, Kristen. *Ham Radio's Technical Culture*. Cambridge, MA: MIT Press, 2006.

Hayes, Joy. *Radio Nation: Communication, Popular Culture, and Nationalism in Mexico, 1920–1945*. Tucson: University of Arizona Press, 2000.

Hazen, Don, and Julie Winokur. *We the Media: A Citizen's Guide to Fighting for Media Democracy*. New York: New Press, 1997.

Herman, Edward, and Noam Chomsky. *Manufacturing Consent: The Political Economy of the Mass Media*. New York: Pantheon Books, 1988.

Herman, Edward, and Robert McChesney. *The Global Media: The New Missionaries of Global Capitalism*. London: Cassell, 1997.

Hess, David. Technology- and Product-Oriented Movements: Approximating Social Movement Studies and Science and Technology Studies. *Science, Technology & Human Values* 30 (2005): 515–535.

Hill, Mike. *Whiteness: A Critical Reader*. New York: New York University Press, 1997.

Himanen, Pekka. *The Hacker Ethic and the Spirit of the Information Age*. New York: Random House, 2001.

Hintz, Arne, and Stefania Milan. Grassroots Tech Activists and Media Policy. In *Encyclopedia of Social Movement Media*, ed. J. Downing, 217–221. Thousand Oaks, CA: Sage, 2010.

Hoffman, Abbie. *Steal This Book*. New York: Four Walls Eight Windows Press. 1996. (Originally published in 1970)

hooks, bell. *Talking Back: Thinking Feminist, Thinking Black*. Cambridge, MA: South End Press, 1989.

Horkheimer, Max, and Theodor Adorno. *The Dialectic of Enlightenment*. New York: Continuum, 1991.

Horowitz, Roger, ed. *Boys and Their Toys? Masculinity, Class, and Technology in America*. New York: Routledge, 2001.

Horwitz, Robert. *The Irony of Regulatory Reform*. New York: Oxford University Press, 1989.

Horwitz, Robert. Broadcast Reform Revisited: Reverend Everett C. Parker and the "Standing" Case (*Office of Communication of the United Church of Christ v. Federal Communications Commission*). *Communication Review* 2 (1997): 311–348.

HoustonIndymedia. Setting up 95.3 KAMP for Broacast [*sic*] Tomorrow (September 12, 2005). Available at http://houston.indymedia.org/news/2005/09/43351.php.

Hoynes, William. Why Media Mergers Matter. Open Democracy website (January 16, 2002). Available at http://www.opendemocracy.net/media-globalmediaownership/article_47.jsp.

Hughes, Everett C. *The Sociological Eye*. Chicago: Aldine Atherton, 1971.

Hughes, Thomas. The Evolution of Large Technological Systems. In *The Social Construction of Technological Systems*, ed. Wiebe Bijker, Thomas P. Hughes, and Trevor J. Pinch. Cambridge, MA: MIT Press, 1987.

Ihde, Don. *Listening and Voice: A Phenomenology of Sound*. Athens: Ohio University Press, 1976.

Jackson, Steven. Rethinking Repair. In *Media Technologies: Essays on Communication, Materiality, and Society*, ed. Tarleton Gillespie, Pablo Boczkowski, and Kirsten Foote, 452–491. Cambridge, MA: MIT Press, 2013.

Janssen, Mike. Intervention by Congress Slashes LPFM Licensing by 80%. *Current* (January 15, 2001a). Available at http://www.current.org/tech/tech0101lpfm.html.

Janssen, Mike. LPFM Rules Still Disputed; Congress May Act. *Current* (October 16, 2001b). Available at http://www.current.org/tech/tech019lpfm.html.

Janssen, Mike. Interference Study Finds Room for More Low-Power FMs. *Current* (August 4, 2003). Available at http://www.current.org/tech/tech0314lpfm.html.

Jasanoff, Sheila. *Science at the Bar*. Cambridge, MA: Harvard University Press, 1995.

Jenkins, Henry. *Convergence Culture*. New York: New York University Press, 2006.

Johns, Adrian. *The Nature of the Book*. Chicago: University of Chicago Press, 1998.

Johns, Adrian. Pop Music Pirate Hunters. *Daedalus* 131 (2002): 67–77.

Jones, Michael O. Why Folklore and Organizations? *Western Folklore* 50 (1991): 29–40.

Jordan, Chris. A Toaster with Pictures: The Political Economy of Reagan Era Entertainment. Presented to Union for Democratic Communications Conference, Eugene, OR (October, 1999).

Jordan, Kathleen, and Michael Lynch. The Sociology of a Genetic Engineering Technique: Ritual and Rationality in the Performance of the "Plasmid Prep." In *The Right Tools for the Job*, ed. Adele Clarke and Joan Fujimura. Princeton, NJ: Princeton University Press, 1992.

Jordan, Tim, and Paul Taylor. A Sociology of Hackers. *The Sociological Review* 46 (2008): 757–780.

Juris, Jeffrey. *Networking Futures: The Movements against Corporate Globalization*. Durham, NC: Duke University Press, 2008.

Karr, Rick. Prometheus Unbound. *Nation* 280 (2005): 22–27.

Keith, Michael. Turn On ... Tune In: The Rise and Demise of Commercial Underground Radio. In *The Radio Studies Reader*, ed. Michele Hilmes and Jason Loviglio. New York: Routledge, 2002.

Kelty, Christopher. Geeks, Social Imaginaries, and Recursive Publics. *Cultural Anthropology* 20 (2005): 185–214.

Kelty, Christopher. *Two Bits: The Cultural Significance of Free Culture*. Durham, NC: Duke University Press, 2008.

Keniston, Kenneth. *Young Radicals: Notes on Committed Youth*. New York: Harcourt Brace Jovanovich, 1968.

Kidd, Dorothy. Talking the Walk: The Media Enclosures and the Communications Commons. PhD diss., Simon Fraser University (1998).

Kidd, Dorothy. Indymedia.org: The Development of the Communication Commons. *Democratic Communiqué* 18 (2002a): 65–86.

Kidd, Dorothy. Which Would You Rather: Seattle or Puerto Alegre? Presented at Our Media, Not Theirs Conference, Barcelona, Spain (July 2002b).

Kirkpatrick, Bill. Localism in American Media Policy, 1920–34: Reconsidering a "Bedrock" Concept. *Radio Journal* 4 (2006): 87–110.

Kleif, Tine, and Wendy Faulkner. "I'm No Athlete [but] I Can Make This Thing Dance!" Men's Pleasures in Technology. *Science, Technology & Human Values* 28 (2003): 296–325.

Kline, Ronald. Agents of Modernity: Home Economists and Rural Electrification, 1925–1950. In *Rethinking Home Economics: Women and the History of a Profession*, ed. Sara Stage and Virginia B. Vincenti. Ithaca, NY: Cornell University Press, 1997.

Kline, Ronald. *Consumers in the Country: Technology and Social Change in Rural America*. Baltimore: Johns Hopkins University Press, 2000.

Kline, Ronald. Resisting Consumer Technology in Rural America: The Telephone and Electrification. In *How Users Matter: The Co-construction of Users and Technology*, ed. Nelly Oudshoorn and Trevor Pinch. Cambridge, MA: MIT Press, 2003.

Kline, Ronald R. What Is Information Theory a Theory Of? Boundary Work among Information Theorists and Information Scientists in the United States and Britain during the Cold War. In *The History and Heritage of Scientific and Technical Information Systems: Proceedings of the 2002 Conference, Chemical Heritage Foundation*, ed. W. Boyd Rayward and Mary Ellen Bowden. Medford, NJ: Information Today, 2004.

Kline, Ronald. Cybernetics, Management Science, and Technology Policy: The Emergence of "Information Technology" as a Keyword, 1948–1985. *Technology and Culture* 47 (2006): 513–535.

Kline, Ronald, and Trevor Pinch. Users as Agents of Technological Change: The Social Construction of the Automobile in the Rural United States. *Technology and Culture* 37 (1996): 763–795.

Klinenberg, Eric. *Fighting for Air: The Battle to Control America's Media*. New York: Metropolitan Books, 2007.

Knorr Cetina, Karin. *Epistemic Cultures: How the Sciences Make Knowledge*. Cambridge, MA: Harvard University Press, 1999.

Kohler, Robert. *Lords of the Fly: Drosophila Genetics and Experimental Life*. Chicago: University of Chicago Press, 1994.

Kreiss, Daniel, Megan Finn, and Fred Turner. The Limits of Peer Production: Some Reminders from Max Weber for the Network Society. *New Media & Society* 13 (2) (2011): 243–259.

Kruse, Holly. *Site and Sound: Understanding Independent Music Scenes*. New York: Peter Lang, 2003.

Kunda, Gideon. *Engineering Culture: Control and Commitment in a High-Tech Corporation*. Philadelphia: Temple University Press, 1992.

Kvasny, Lynnette. The Role of the Habitus in Shaping Discourses about the Digital Divide. *Journal of Computer-Mediated Communication* 10 (2): article 5 (2005). Available at http://jcmc.indiana.edu/vol10/issue2/kvasny.html.

Lagemann, Ellen Condliffe, ed. *Philanthropic Foundations: New Scholarship, New Possibilities*. Bloomington: Indiana University Press, 1999.

Lasar, Matthew. *Pacifica Radio: The Rise of an Alternative Network*. Philadelphia: Temple University Press, 2000.

Latour, Bruno. Visualization and Cognition: Thinking with Eyes and Hands. *Knowledge and Society: Studies in the Sociology of Culture Past and Present* 6 (1986): 1–40.

Latour, Bruno. *Science in Action*. Cambridge, MA: Harvard University Press, 1987.

Latour, Bruno. *The Pasteurization of France*. Cambridge, MA: Harvard University Press, 1988.

Latour, Bruno, and Steve Woolgar. *Laboratory Life: The Social Construction of Scientific Facts*. Princeton, NJ: Princeton University Press, 1979.

Leonhardt, David. Religious Groups Are Pushing Hard to Go on the Air with Low-Power Radio Stations. *New York Times* (July 11, 2000). Available at http://www.nytimes.com/financial/071100radio-fcc-html.

Lerman, Nina, Arwen Palmer Mohun, and Ruth Oldenziel. Versatile Tools: Gender Analysis and the History of Technology. *Technology and Culture* 38 (1997): 1–9.

Lerman, Nina E., Arwen Palmer Mohun, and Ruth Oldenziel. *Gender & Technology: A Reader*. Baltimore: Johns Hopkins University Press, 2003.

Lessig, Lawrence. *Code and Other Laws of Cyberspace*. New York: Basic Books, 1999.

Levy, Steven. *Hackers: Heroes of the Computer Revolution*. New York: Penguin, 1984.

Light, Jennifer. When Computers Were Women. *Technology and Culture* 40 (1999): 455–483.

Lindsay, Christina. From the Shadows: Users as Designers, Producers, Marketers, Distributors and Technical Support. In *How Users Matter: The Co-construction of Users and Technology*, ed. Nelly Oudshoorn and Trevor Pinch. Cambridge, MA: MIT Press, 2003.

Lloyd, Mark. Lessons for Realistic Radicals in the Information Age. In *The Future of Media: Resistance and Reform in the 21st Century*, ed. Robert McChesney, Russell Newman, Ben Scott, and Bill Moyers. New York: Seven Stories, 2005.

Lullabot Ideas. Drupal Voices 100: Jack Aponte on Diversity, Power and Privilege in Open Source Communities (May 12, 2010). Available at http://www.lullabot.com/podcasts/drupal-voices-100-jack-aponte-on-diversity-power-and-privilege-open-source-communities.

MacKenzie, Donald. Theories of Technology and the Abolition of Nuclear Weapons. In *The Social Shaping of Technology*. 2nd ed., ed. Donald MacKenzie and Judy Wajcman. Buckingham, UK: Open University Press, 1999.

Marvin, Carolyn. *When Old Technologies Were New: Thinking about Electric Communication in the Late Nineteenth Century*. New York: Oxford University Press, 1988.

Marx, Leo. Technology: The Emergence of a Hazardous Concept. *Technology and Culture* 51 (2010): 561–577.

McAdam, Douglas, Sidney Tarrow, and Charles Tilly, eds. *Dynamics of Contention*. New York: Cambridge University Press, 2001.

McChesney, Robert. *Rich Media, Poor Democracy*. New York: The New Press, 1999.

McChesney, Robert. Media Policy Goes to Main Street: The Uprising of 2003. *Communication Review* 7 (2004): 223–258.

McChesney, Robert. The Emerging Struggle for a Free Press. In *The Future of Media: Resistance and Reform in the 21st Century*, ed. Robert McChesney, Russell Newman, Ben Scott, and Bill Moyers. New York: Seven Stories, 2005.

McChesney, Robert. FCC Scandal Explodes with Second Revelation of Suppressed Media Ownership Research. CommonDreams website (September 19, 2006). Available at http://www.commondreams.org/views06/0919-27.htm.

McChesney, Robert, Russell Newman, Ben Scott, and Bill Moyers, eds. *The Future of Media: Resistance and Reform in the 21st Century*. New York: Seven Stories, 2005.

McCaughey, Martha, and Michael Ayers, eds. *Cyberactivism: Online Activism in Theory and Practice*. New York: Routledge, 2003.

McCauley, Michael. Radio's Digital Future: Preserving the Public Interest in the Age of New Media. In *The Radio Studies Reader*, ed. Michele Hilmes and Jason Loviglio. New York: Routledge, 2002.

McLaughlin, Lisa, and Victor Pickard. What Is Bottom-Up about Global Internet Governance? *Global Media and Communication* 1 (2005): 357–373.

McLuhan, Marshall. *Understanding Media: The Extensions of Man*. New York: Mentor Books, 1964.

Mediageek website. Be the Media: Blogging from the Audience, Hallways and Streets of the 2005 National Conference for Media (May 14, 2005). Available at http://www.mediageek.org/btm/archives/2005/05/indymedia_caucu.html.

Meinrath, Sascha. Wirelessing the World: The Battle over (Community) Wireless Networks. In *The Future of Media: Resistance and Reform in the 21st Century*, ed. Robert McChesney, Russell Newman, Ben Scott, and Bill Moyers. New York: Seven Stories, 2005.

Microradio Implementation Project website (2005). Available at http://www.microradio.org/mr021102.htm.

Melucci, Alberto. *Challenging Codes: Collective Action in the Information Age*. New York: Cambridge University Press, 1996.

Milam, Lorenzo W. *Sex and Broadcasting: A Handbook on Starting a Radio Station for the Community*. Saratoga, CA: Dildo Press, 1975.

Misa, Thomas J. *A Nation of Steel: The Making of Modern America, 1865–1925*. Baltimore: Johns Hopkins University Press, 1995.

MITRE Corporation. Experimental Measurements of the Third-Adjacent Channel Impacts of Low-Power FM Stations. Sponsored by the Federal Communications Commission 1201FCC3-AA Contract No. CON01000020 (May 2003).

Mody, Cyrus C.M. Crafting the Tools of Knowledge: The Invention, Spread, and Commercialization of Probe Microscopy, 1960–2000. PhD diss., Cornell University (2004).

Mueller, M. Interest Groups and the Public Interest: Civil Society Action and the Globalization of Communications Policy. Paper presented to the Telecommunications Policy Research Conference, Arlington, VA (September 11, 2002).

Napoli, Philip. Public Interest Media Advocacy and Activism as a Social Movement. In *Communication Yearbook 33*, ed. C. Beck. New York: Routledge, 2009.

National Association of Broadcasters. Comments of the National Association of Broadcasters on the MITRE Corporation Report. Federal Communications Commission, in the Matter of Creation of Low Power Radio Service, MM Docket No. 99–25 (October 13, 2003).

National Association of Broadcasters. Legislative Issue Paper: Low-Power FM (LPFM) (September 2004). Available at http://www.nab.org/newsroom/issues/issuepapers/issueslpfm.asp.

National Public Radio website. About NPR. Available at http://www.npr.org/about/.

National Public Radio. Statement of National Public Radio, Inc. Federal Communications Commission, Proposal for Creation of the Low-Power FM (LPFM) Broadcast Service, RM-9242 (April 27, 1998).

National Public Radio. NPR Supports HR 3439, The Radio Broadcasting Preservation Act of 2000. Press release (March 30, 2000). Available at http://www.npr.org/about/press/000406.lpfmlegislation.html.

Neff, Gina, and David Stark. Permanently Beta: Responsive Organization in the Internet Era. In *Society Online*, ed. Philip Howard and Steve Jones. Thousand Oaks, CA: Sage, 2004.

Negroponte, Nicholas. *Being Digital*. New York: Vintage Books, 1995.

New Orleans Indymedia. Creativity, Solidarity, and Mutual Aid in Algiers, New Orleans (September 13, 2005). Available at http://neworleans.indymedia.org/news/2005/09/5158.php.

New York State Attorney General. Sony Settles Payola Investigation. Press release (July 25, 2005). Available at http://www.oag.state.ny.us/press/2005/jul/jul25a_05.html.

Nielsen, Rasmus Kleis. The Labors of Internet-Assisted Activism: Overcommunication, Miscommunication, and Communicative Overload. *Journal of Information Technology & Politics* 6 (2009): 267–280.

Office of the New York State Attorney General. Sony Settles Payola Investigation (July 25, 2005).

Oldenziel, Ruth. Boys and Their Toys: The Fisher Body Craftsman's Guild, 1930–1968, and the Making of a Male Technical Domain. *Technology and Culture* 38 (1997): 60–96.

Oldenziel, Ruth. *Making Technology Masculine: Men, Women, and Modern Machines in America, 1870–1945*. Amsterdam: Amsterdam University Press, 2004.

Opel, Andy. *Micro Radio and the FCC: Media Activism and the Struggle over Broadcast Policy*. Westport, CT: Praeger, 2004.

Orlikowski, Wanda. Using Technology and Constituting Structures: A Practice Lens for Studying Technology in Organizations. *Organization Science* 11 (2000): 404–428.

Orlikowski, Wanda. Sociomaterial Practices: Exploring Technology at Work. *Organization Studies* 28 (2007): 1435–1448.

Orr, Julian. Sharing Knowledge, Celebrating Identity: Community Memory in a Service Culture. In *Collective Remembering*, ed. David Middleton and Derek Edwards. London: Sage, 1990.

Orr, Julian. *Talking about Machines: An Ethnography of a Modern Job*. Ithaca, NY: Cornell University Press, 1996.

Oudshoorn, Nelly, and Trevor Pinch, eds. *How Users Matter: The Co-construction of Users and Technology*. Cambridge, MA: MIT Press, 2003.

Oudshoorn, Nelly, Els Rommes, and Marcelle Stienstra. Configuring the User as Everybody: Gender and Design Cultures in Information and Communication Technologies. *Science, Technology & Human Values* 29 (2004): 30–63.

Parthasarathy, Shobita. Knowledge Is Power: Genetic Testing for Breast Cancer and Patient Activism in the United States and Britain. In *How Users Matter: The Co-Construction of Users and Technology*, ed. Nelly Oudshoorn and Trevor Pinch. Cambridge, MA: MIT Press, 2003.

Pfaffenberger, Bryan. The Social Meaning of the Personal Computer: Or, Why the Personal Computer Revolution Was No Revolution. *Anthropological Quarterly* 61 (1988): 39–47.

PhillyIMC website. About. Available at http://phillyimc.org/en/static/how.shtml.

Pickard, Victor. United Yet Autonomous: Indymedia and the Struggle to Sustain a Radical Democratic Network. *Media Culture & Society* 28 (2006): 315–336.

Pickard, Victor. Cooptation and Cooperation: Institutional Exemplars of Democratic Internet Technology. *New Media & Society* 10 (2008): 625–645.

Pickard, Victor. The Battle over the FCC Blue Book: Determining the Role of Broadcast Media in a Democratic Society, 1945–8. *Media Culture & Society* 33 (2) (2011): 171–191.

Pickard, Victor. *America's Battle for Media Democracy*. Cambridge: Cambridge University Press, 2014.

Piette, Betsy. Radio Mutiny Hijacked in Philadelphia. *Workers World*. July 23, 1998. http://www.workers.org/ww/1998/pirate0723.php.

Pinch, Trevor. Why You Go to a Piano Store to Buy a Synthesizer: Path Dependence and the Social Construction of Technology. In *Path Dependence and Creation*, ed. Raghu Garud and Peter Karnoe. Mahwah, NJ: Lawrence Erlbaum, 2001.

Pinch, Trevor, and Wiebe Bijker. The Social Construction of Facts and Artifacts: Or How the Sociology of Science and the Sociology of Technology Might Benefit Each Other. In *the Social Construction of Technological Systems*, ed. Wiebe Bijker, Thomas P. Hughes, and Trevor J. Pinch. Cambridge, MA: MIT Press, 1987.

Pinch, Trevor, and Frank Trocco. *Analog Days: The Invention and Impact of the Moog Synthesizer*. Cambridge, MA: Harvard University Press, 2002.

Polletta, Francesca. *Freedom Is an Endless Meeting*. Chicago: University of Chicago Press, 2004.

Polletta, Francesca. *It Was Like a Fever: Storytelling in Protest and Politics*. Chicago: University of Chicago Press, 2006.

Pooley, Jefferson. From Psychological Warfare to Social Justice. In *Media and Social Justice*, ed. Sue Curry Jansen, Jeff Pooley, and Lora Taub-Pervizpour. New York: Palgrave MacMillan, 2011.

Porter, Theodore. *Trust in Numbers: The Pursuit of Objectivity in Science and Public Life*. Princeton, NJ: Princeton University Press, 1996.

Portwood-Stacer, Laura. *Lifestyle Politics and Radical Activism*. New York: Continuum Press, 2013.

Postigo, Hector. From Pong to Planet Quake: Post-industrial Transitions from Leisure to Work. *Information Communication and Society* 6 (4) (2003): 593–607.

Prometheus Radio Project. Comments of the Prometheus Radio Project before the Federal Communications Commission. MM Docket No. 99–25 (July 29, 1999).

Prometheus Radio Project. Radio Free Nashville Unbuckles the Bible Belt's Airwaves. Press release (March 28, 2005).

Prometheus Radio Project Website. Our Third Barnraising—KOCZ-LP in Opelousas, Louisiana. Available at http://www.prometheusradio.org/KOCZ_barnraising.

Pursell, Carroll. The Rise and Fall of the Appropriate Technology Movement in the United States, 1965–1985. *Technology and Culture* 34 (1993): 629–637.

Rabinow, Paul. *French DNA: Trouble in Purgatory*. Chicago: University of Chicago Press, 1999.

RECNET. LPFM Scoreboard. Available at http://recnet.net/scoreboard.php?wind=1.

Riles, Annelise. *The Network Inside Out*. Ann Arbor: University of Michigan Press, 2000.

Riismandel, Paul. Radio by and for the Public. In *The Radio Studies Reader*, ed. Michele Hilmes and Jason Loviglio. New York: Routledge, 2002.

Rose, Dale, and Stuart Blume. Citizens as Users of Technology: An Exploratory Study of Vaccines and Vaccination. In *How Users Matter: The Co-construction of Users and Technology*, ed. Nelly Oudshoorn and Trevor Pinch. Cambridge, MA: MIT Press, 2003.

Rosen, Paul. The Social Construction of Mountain Bikes: Technology and Postmodernity in the Cycle Industry. *Social Studies of Science* 23 (1993): 479–513.

Rosenzweig, Roy. Can History Be Open Source? *Wikipedia* and the Future of the Past. *Journal of American History* 93 (2006): 117–146.

Rothenbuhler, Eric, and Tom McCourt. Radio Redefines Itself, 1947–1962. In *The Radio Studies Reader*, ed. Michele Hilmes and Jason Loviglio. New York: Routledge, 2002.

Ruggiero, Greg. *Microradio & Democracy: (Low) Power to the People*. New York: Seven Stories Press, 1999.

Sandvig, Christian. The Return of the Broadcast War. Presented to the 33rd Research Conference on Communication, Information, and Internet Policy, Arlington, VA (September 23, 2005).

Sandvig, Christian, David Young, and Sascha Meinrath. Hidden Interfaces to "Ownerless" Networks. Presented to the 32nd Research Conference on Communication, Information, and Internet Policy, Arlington, VA (September 2004).

Schiller, Dan. How to Think about Information. In *The Political Economy of Information*, ed. Vincent Mosco and Janet Wasko. Madison: University of Wisconsin Press, 1988.

Schiller, J. Zach. On Becoming the Media: Low Power FM and the Alternative Public Sphere. In *Media and Public Spheres*, ed. Richard Butsch. New York: Palgrave Macmillan, 2007.

Scholz, Trebor. *Digital Labor: The Internet as Playground and Factory*. New York: Routledge, 2012.

Schutz, Alfred. Common Sense and Scientific Interpretation of Human Action. In *Collected Papers I: The Problem of Social Reality,* ed. Maurice Natanson. The Hague, The Netherlands: Martinus Nijhoff, 1973.

Scott, Pam, Evelleen Richards, and Brian Martin. Captives of Controversy: The Myth of the Neutral Social Researcher in Contemporary Scientific Controversies. *Science, Technology & Human Values* 15 (1990): 474–494.

Shannon, Claude. A Mathematical Theory of Communication. *Bell System Technical Journal* 27 (1948): 379–423.

Shannon, Claude, and Warren Weaver. *The Mathematical Theory of Communication.* Urbana: University of Illinois Press, 1949.

Shields, Steven O., and Robert Ogles. Black Liberation Radio: A Case Study of Free Radio Micro-broadcasting. *Howard Journal of Communications* 5 (1995): 173–183.

Shirky, Clay. *Here Comes Everybody*. New York: Penguin Press, 2008.

Slotten, Hugh. *Radio and Television Regulation: Broadcast Technology in the United States, 1920–1960*. Baltimore: Johns Hopkins University Press, 2000.

Smith, Merritt Roe. Technological Determinism in American Culture. In *Does Technology Drive History? The Dilemma of Technological Determinism*, ed. Merritt Roe Smith and Leo Marx. Cambridge, MA: MIT Press, 1994.

Smulyan, Susan. *Selling Radio: The Commercialization of American Broadcasting, 1920–1934*. Washington, DC: Smithsonian Institution Press, 1994.

Social Science Research Council. SSRC Receives $1.5 Million Ford Grant to Continue Innovative Work on Media Reform (June 22, 2007). Available at http://www.ssrc.org/press/fordgrant062107/.

Söderberg, Johan. Free Space Optics in the Czech Wireless Community: Shedding Some Light on the Role of Normativity for User-Initiated Innovations. *Science, Technology & Human Values* 36 (4) (2011): 423–450.

Soley, Lawrence. *Free Radio: Electronic Civil Disobedience*. Boulder, CO: Westview Press, 1998.

Starr, Paul. *The Creation of the Media*. New York: Basic Books, 2004.

Stein, Arlene. Sex, Truths, and Audiotape: Anonymity and the Ethics of Exposure in Public Ethnography. *Journal of Contemporary Ethnography* 39 (2010): 554–568.

Steinberg, Jacques. Facing Criticism, F.C.C. Is Thinking Local. *New York Times* (August 20, 2003).

Sterne, Jonathan. The MP3 as Cultural Artifact. *New Media & Society* 8 (2006): 825–842.

Straus, Scott. What Is the Relationship between Hate Radio and Violence? Rethinking Rwanda's "Radio Machete." *Politics & Society* 35 (2007): 609–637.

Streeter, Thomas. *Selling the Air: A Critique of the Policy of Commercial Broadcasting in the United States*. Chicago: University of Chicago Press, 1996.

Streeter, Thomas. *The Net Effect: Romanticism, Capitalism, and the Internet*. New York: New York University Press, 2011.

Sturken, Marita, and Douglas Thomas. Introduction. In *Technological Visions: The Hopes and Fears That Shape New Technologies*, ed. Marita Sturken, Douglas Thomas, and Sandra Ball-Rokeach. Philadelphia: Temple University Press, 2004.

Suchman, Lucy. *Plans and Situated Actions: The Problem of Human-Machine Communication*. New York: Cambridge University Press, 1987.

Suchman, Lucy. *Located Accountabilities in Technology Production*. Lancaster: Centre for Science Studies, Lancaster University, 2003. Available at http://www.comp.lancs.ac.uk/sociology/papers/Suchman-Located-Accountabilities.pdf.

Takahashi, Yuzo. A Network of Tinkerers: The Advent of the Radio and Television Receiver Industry in Japan. *Technology and Culture* 41 (2000): 460–484.

Tapia, Andrea, C. Maitland, and M. Stone. Making IT Work for Municipalities: Building Municipal Wireless Networks. *Government Information Quarterly* 23 (2006): 359–380.

Tarrow, Sidney. *Power in Movement: Social Movements, Collective Action and Politics*. Cambridge, UK: Cambridge University Press, 1998.

Terkel, Studs. *Working*. New York: The New Press, 1972.

Terranova, Tiziana. *Network Culture: Politics for the Information Age*. Ann Arbor, MI: Pluto Press, 2004.

Théberge, Paul. *Any Sound You Can Imagine: Making Music/Consuming Technology*. Hanover, NH: University Press of New England, 1997.

Thompson, E. P. *The Making of the English Working Class*. Harmondsworth, UK: Penguin, 1968.

Tönnies, Ferdinand. *Community and Civil Society*, ed. Jose Harris. Cambridge, UK: Cambridge University Press, 2001.

Traweek, Sharon. *Beamtimes and Lifetimes: The World of High Energy Physicists*. Cambridge, MA: Harvard University Press, 1988.

Turkle, Sherry. *The Second Self: Computers and the Human Spirit*. New York: Simon & Schuster, 1984.

Turkle, Sherry. *Life on the Screen: Identity in the Age of the Internet.* New York: Simon & Schuster, 1995.

Turner, Fred. *From Counterculture to Cyberculture.* Chicago: University of Chicago Press, 2006.

Turner, Fred. Burning Man at Google: A Cultural Infrastructure for New Media Production. *New Media & Society* 11 (2009): 73–94.

Turner, Victor. *The Ritual Process: Structure and Anti-Structure.* Chicago: Aldine, 1969.

Turner, Victor. *The Anthropology of Performance.* New York: PAJ Publications, 1987.

United Church of Christ et al., filed by Media Access Project. Comments of United Church of Christ, Office of Communication et al. before the Federal Communications Commission. MM Docket No. 99–25 (August 2, 1999).

United States Communications Act of 1934 as amended by the 1996 Telecommunications Act. Approved by the 104th Congress on January 3, 1996.

United States Office of the Federal Register. Code of Federal Regulations.

United States Public Broadcasting Act of 1967 (47 U.S.C. § 396).

United States Senate. Senator John McCain, Remarks on the Floor of the Senate (October 26, 2000). Available at http://prometheusradio.org/content/view/86/134.

United States Senate. 109th Congress, 1st Session. Local Community Radio Act of 2005 (S-312) (February 8, 2005).

Vaillant, Derek. "Your Voice Came in Last Night, but I Thought It Sounded a Little Scared": Rural Radio Listening and "Talking Back" during the Progressive Era in Wisconsin, 1920–1932. In *The Radio Studies Reader*, ed. Michele Hilmes and Jason Loviglio. New York: Routledge, 2002.

Van Maanen, John, and Stephen Barley. Occupational Communities: Culture and Control in Organizations. In *Research in Organizational Behavior.* vol. 6, ed. Barry Staw and L. L. Cummings, 287–365. Greenwich, CT: JAI Press, 1984.

Waits, Jennifer. Radio Still Growing in U.S. According to Latest FCC Stats. RadioSurvivor.org (January 6, 2012). Available at http://www.radiosurvivor.com/2012/01/06/radio-still-growing-in-u-s-according-to-latest-fcc-stats.

Wajcman, Judy. *Feminism Confronts Technology.* University Park: Pennsylvania State University Press, 1991.

Waksman, Steve. California Noise: Tinkering with Hardcore and Heavy Metal in Southern California. *Social Studies of Science* 34 (2004): 675–702.

Walker, Jesse. *Rebels on the Air.* New York: New York University Press, 2001.

Wall, Derek. *Earth First! and the Anti-Roads Movement.* London: Routledge, 1999.

Wang, Jennifer. "The Case of the Radio-Active Housewife": Relocating Radio in the Age of Television. In *The Radio Studies Reader*, ed. Michele Hilmes and Jason Loviglio. New York: Routledge, 2002.

Weber, Max. *The Theory of Social and Economic Organization,* trans. A. M. Henderson and Talcott Parsons. London: Collier Macmillan, 1947.

Webster, Frank. *Theories of the Information Society*. 2nd ed. New York: Routledge, 2002.

Wells, Jamese. *Spirit of Resistance*. Independent film. 2006.

Wikipedia. Pirate Radio. Available at http://en.wikipedia.org/wiki/Free_radio.

Williams, Raymond. *Keywords*. New York: Oxford University Press, 1976.

Winner, Langdon. *The Whale and the Reactor*. Chicago: University of Chicago Press, 1988.

WJFF Radio Catskill website. The Story of Hydropowered Radio. Available at http://www.wjffradio.org/wjff/index.php?section=19.

Wolfson, Todd. Mediated Democracy: Activism and the Promise of Politics in Cyber Capitalism. Presented to American Anthropological Association, Washington, DC (December 2005).

Wolfson, Todd. The Cyber Left: Indymedia and the Making of 21st-Century Social Movements. PhD diss., University of Pennsylvania (2008).

Wolfson, Todd. From the Zapatistas to Indymedia: Dialectics and Orthodoxy in Contemporary Social Movements. *Communication, Culture & Critique* 5 (2012): 149–170.

Woodhouse, Edward, David Hess, Steve Breyman, and Brian Martin. Science Studies and Activism: Possibilities and Problems for Reconstructivist Agendas. *Social Studies of Science* 32 (2002): 297–319.

Woolgar, Steve. Configuring the User: The Case of Usability Trials. In *A Sociology of Monsters: Essays on Power, Technology and Domination*, ed. John Law. New York: Routledge, 1991.

Wyatt, Sally. Non-users Also Matter. In *How Users Matter: The Co-construction of Users and Technology*, ed. Nelly Oudshoorn and Trevor Pinch. Cambridge, MA: MIT Press, 2003.

Wynne, Brian. Misunderstood Misunderstandings: Social Identities and the Public Uptake of Science. In *Misunderstanding Science? The Public Reconstruction of Science and Technology*, ed. Alan Irwin and Brian Wynne. Cambridge, UK: Cambridge University Press, 1996.

Index

Italic page numbers indicate illustrations.

Printed in the United States
By Bookmasters